THE ULTIMATE DIY

GEEK TOYS GUIDE

EDITED BY DOUG CANTOR

POPULAR SCIENCE THE FUTURE NOW

THE ULTIMATE DIY GEEK TOYS GUIDE

weldonowen

contents

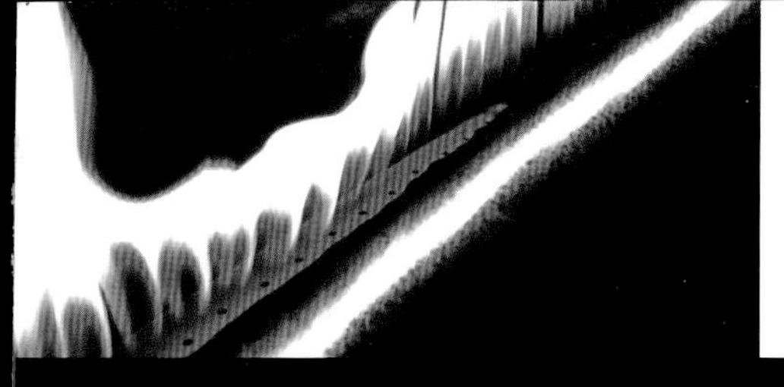

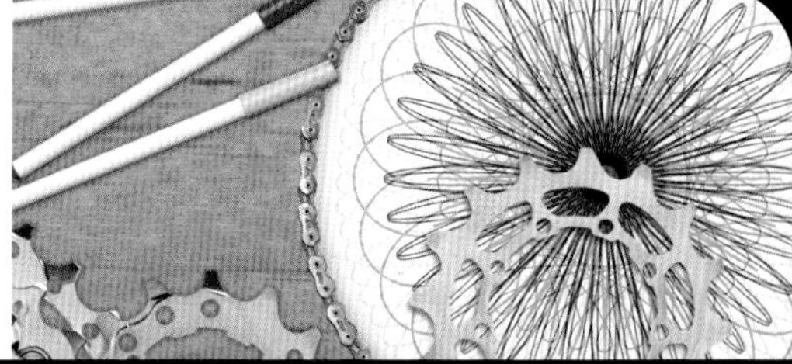

FOREWORD

Anyone can make anything. That is the lesson of the 140 years that *Popular Science* has been in print. A determined person, working on the weekends, can escape gravity, break the speed of sound, or create a new means of communicating across great distances. And that inventive process begins by tearing things apart and rebuilding them again.

I cannot claim to have been the kind of kid who did that. My instincts around technology were always to keep my belongings clean, dry, and otherwise in perfect working order, not to dismantle or modify them. But my years at *Popular Science* have taught me that that instinct is the wrong one. There is simply too much technology at our disposal not to mess around with it. And when a determined person brings his or her inventive instincts to bear on the gadgets and gizmos that fill our lives, great things can result.

But this book isn't necessarily about building great things. It's about messing around, usually for the simple fun of it. The projects on these pages are based, in part, on the years we've spent pursuing the lone, sometimes crazed hackers who don't just modify technology but blow it apart, just to be able to say they've done it.

I spent an afternoon with our staff photographer, John Carnett, who in his off time had replaced the motor on a four-wheel ATV with a jet engine. The thing required an elaborate start-up procedure and ear protection just to get it rolling, and as he drove me through his Philadelphia neighborhood in it, I cringed to imagine the incredible racket we were making, essentially that of a 747 taxiing past. And yet throughout the process, oblivious to the enemies he was making among the local parents, dogs, and nappers, John wore a look of joy and pride that had

nothing to do with serving humanity or inventing something new. He'd just tricked something out—hacked it—in his own way, and in doing so had made his mark on the universe, not to mention on the neighborhood noise ordinances.

It's in that spirit that we, and especially our tireless senior editor Doug Cantor, bring you this entertaining collection of projects. We did it for the hell of it.

Jacob Ward

Jacob Ward
Editor-in-Chief
Popular Science

INTRODUCTION

I was an unlikely candidate to become the editor of How 2.0, *Popular Science*'s do-it-yourself column. I've always been reasonably handy, but when it came to real hands-dirty, open-things-up-and-rearrange-the-parts hacking skills, I was a complete novice.

So I got into the DIY world the same way an experienced DIYer would work on a project: I did some research, talked to seasoned tinkerers, and then dove in. Early on I managed to build a tiny flashlight, hack my cell phone's firmware, and make a pair of bookends from old CDs, all without causing too much damage. Over time I found that with a few hours, a small pile of parts from Radio Shack, and a little patience, I could build some really cool stuff.

Editing How 2.0 has also given me a window into the vast community of smart, dedicated people who have found about a million uses for things like solenoid valves and Arduino microcontrollers. The breadth of their innovations is truly astonishing, and that's what *Popular Science* has tried to show in the pages of How 2.0 each month. We've featured projects ranging from a remote-controlled helicopter only 2 inches (5 cm) high to a remote-controlled bomber with a 20-foot (60-m) wingspan; from a portable solar-powered gadget charger to a 200-pound (90-kg) solar-powered 3-D printer; from a robot built from a toothbrush head to one that can mix and serve cocktails.

The projects in this book are the funnest of the fun hacks we've come across, selected from *The Big Book of Hacks*—and they're guaranteed to upgrade your entertainment level. A few of them are just amazing, audacious things that almost no one could (or probably should) try. Most, though, are easier to replicate. Some take only a few minutes and require little more than gluing parts together.

So if you've never attempted to make anything before in your life, this book provides plenty of ways to start. From there you can take on some of the more challenging projects and develop new skills. Eventually, you might actually find yourself advancing your projects far beyond the versions in the book.

Whatever your skill level and area of interest, I encourage you to roll up your sleeves and (safely) give one of these projects a shot. At times you may get frustrated or even break something. But ultimately you'll be surprised by what you can make, hack, tweak, improve, and transform—and by how much fun you'll have doing it.

Douglas Cantor

Doug Cantor
Senior Editor
Popular Science

HOW TO USE THIS BOOK

So you want to hack stuff—to tear it apart, put it back together with other components, and make it new. We at *Popular Science* salute you, and we've put together these projects to get you started. Many of them come from our popular How 2.0 column, and many come from amazingly inventive individuals out there making cool stuff. (Check out the "Thanks to Our Makers" section for more info.) Before starting a project, you can look to the following symbols to decode what you're facing.

If you're just breaking in your screwdriver and have never even heard the word *microcontroller*, try out these projects first. Designed to be doable within five minutes—give or take a few seconds, depending on your dexterity—and to make use of basic household items, these tech crafts are the perfect starting ground for the newbie tinkerer.

Popular Science has been doing DIY for a very long time—almost as long as the 140 years the magazine has been in print. Occasionally this book shares a DIY project from our archives so you can try your hand at the hilarious retro projects your grandfather and grandmother built back before, say, television or smartphones.

Everyone loves a good success story—tales of everyday individuals who created something so wild that it makes you say . . . well, "You built WHAT?!" You'll find several of these stories throughout the book, and it is our hope that they will inspire you to take your projects to the next level.

These are the big ones—the ambitious projects that you'll want to sink some real time and cash into, and that will challenge your skills as a builder. How much time and cash, you ask? And just how challenging? The helpful rubric below will give you an idea.

COST

$ = UNDER $50

$$ = $50–$300

$$ = $300–$1,000

$$$$ = $1,000 AND UP

TIME

◷	UNDER 1 HOUR
◷ ◷	2–5 HOURS
◷ ◷ ◷	5–10 HOURS
◷ ◷ ◷ ◷	10 HOURS AND UP

DIFFICULTY

● ○ ○ ○ ○ One step up from 5 Minute Projects, these tutorials require basic builder smarts, but no electronics or coding wizardry.

● ● ○ ○ ○ Slightly advanced building skills are a must for these projects. Turn on your common sense, and troubleshoot as you go—it's part of the fun.

● ● ● ○ ○ If you see three dots, it means an activity demands low-level electronics and coding skills, or it's pretty rigorous, construction-wise.

● ● ● ● ○ You likely need serious circuitry know-how and code comprehension to do these projects. That, or be prepared to sweat with heavy-duty assembly.

● ● ● ● ● Hey, if you're reading this book, we figure you like challenges. And projects marked with five dots are sure to deliver just that.

WARNING

If you see this symbol, we mean business. Several of the projects in this book involve dangerous tools, electrical current, flammables, potentially harmful chemicals, and recreational devices that could cause injury if misused. So remember: With great DIY comes great responsibility. Use your head, know your tools (and your limitations), always wear safety gear, and never employ your hacking prowess to hurt others. (See our Disclaimer for more information about how *Popular Science* and the publisher are not liable for any mishaps.)

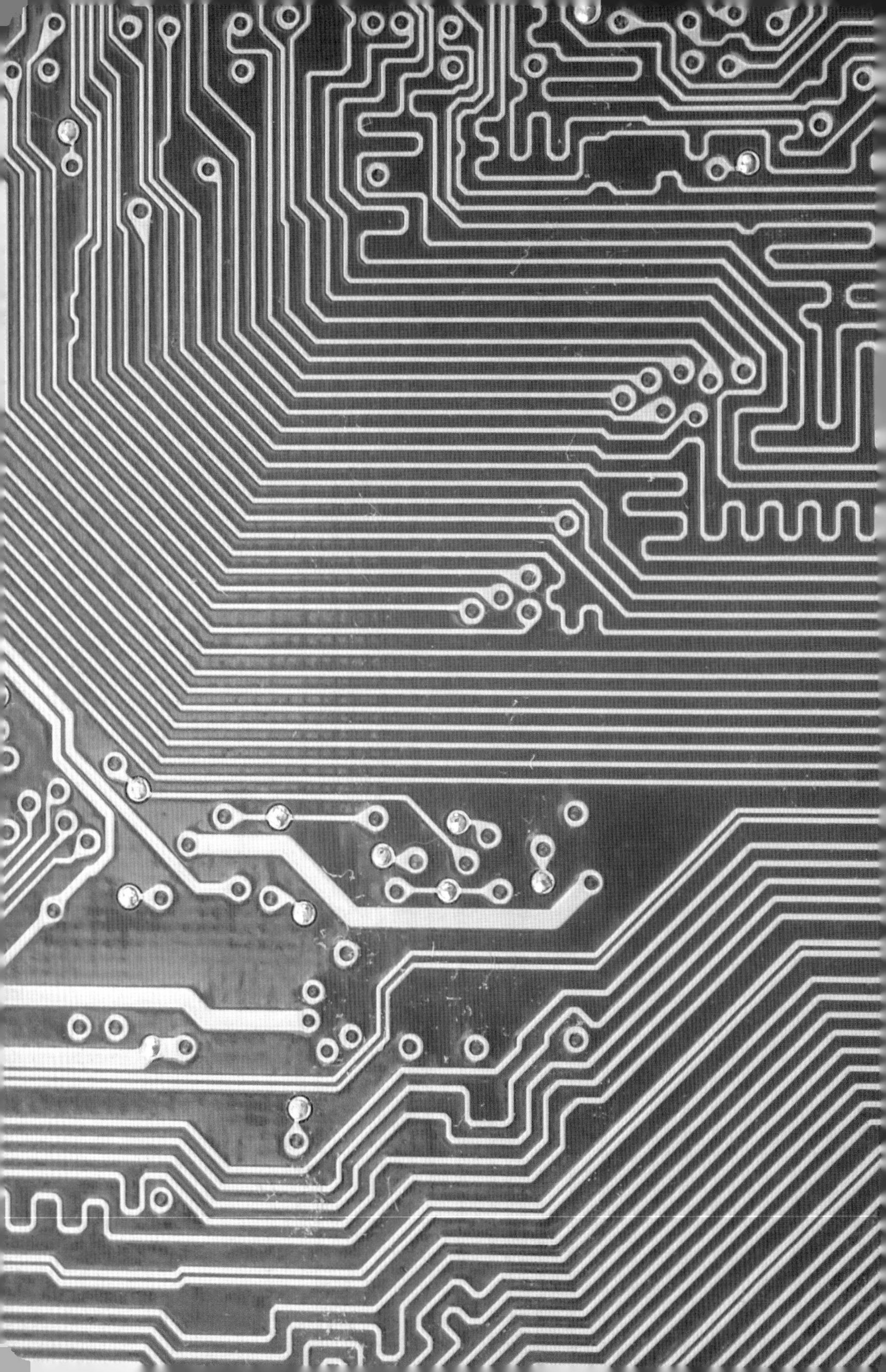

BASIC BUILDING SMARTS

001 PUT TOGETHER A SOLDERING KIT

Soldering is playing with fire, or at least with hot metal. So you need the right tools.

If you're working on electronics projects, you'll need to connect metal objects with a strong, conductive bond. Soldering is the way to go. You heat pieces of metal with a soldering iron, then join them together using a molten filler, or solder.

SOLDER This is the good stuff—the material you'll melt to connect metals. Traditionally, solder was a mix of tin and lead, but these days look for lead-free types to avoid nasty health risks. Choose thinner solder for delicate projects, like attaching wires to a circuit board, and thicker solder for projects involving heftier wires or bulkier pieces of metal.

SOLDERING IRON This tool has a metal tip and an insulated handle. When it's powered on, the tip heats up so it can melt solder. There are low- and high-wattage versions: Low wattage is useful for fragile projects, while high wattage is better for projects involving bigger pieces. There are also different types of tips available for the soldering iron.

SOLDERING IRON STAND Buy a stand that fits your iron so you'll have a place to put it down safely when it's hot. (Leaving this thing lying around when it's turned on is a good way to burn down the toolshed before you've even made anything cool with it!)

CLIPS AND CLAMPS Soldering requires both hands, so you'll need something to hold the materials you're soldering in place. Clips, clamps, and even electrical tape can do the job.

WIRE-MODIFYING TOOLS You'll likely be soldering a lot of wire, so it's useful to have wire cutters, wire strippers, and needle-nose pliers on hand so you can manipulate the wire. Before connecting wires, you must peel back their insulation to expose the wires, so wire strippers are definitely a must.

LIQUID FLUX Soldering works best when the items being soldered are squeaky clean, so have liquid flux on hand—it chases away oxides and other goop that can make soldering difficult.

TIP CLEANER Your soldering iron's tip will get a bit nasty as you work, so keep a wet sponge on hand to periodically wipe down the tip.

HEAT-SHRINK TUBING You can use plastic heat-shrink tubing to insulate wires before you apply heat and solder them. It's available in several diameters for projects with various wire sizes.

EXHAUST FAN The fumes from soldering are not healthy to breathe, so you need good ventilation from a fan or an open window to help clear the air.

SAFETY GOGGLES Bits of hot solder can go flying as you work, so don't do it without wearing safety goggles.

002 LEARN SOLDERING BASICS

Now that you have your soldering gear together, here's how to get it done.

At its most basic, soldering is simply attaching wires to wires. But when soldering onto a circuit board, the process is a little different.

SOLDERING WIRES

STEP 1 In a well-ventilated space, with your safety goggles on, plug in your soldering iron to heat it up.

STEP 2 Prepare the materials you want to join with solder. If you're connecting two wires, peel back any insulation about 1/2 inch (1.25 cm), and twist the wires together. Place your materials on a work surface like scrap wood.

STEP 3 Cut a length from the spool of solder and coil it up at one end, leaving a short lead. You can hold on to the coiled end as you apply the solder.

STEP 4 Touch the iron to the point where the wires are twisted together. Leave it there until the wires are hot enough to melt the solder (about 10 seconds), then touch the solder to the wire joint every few seconds until it begins to melt. Allow enough solder to melt onto the wires to cover them, then pull the solder and soldering iron away. Don't touch the solder directly to the soldering iron during this process.

STEP 5 To fix a mistake, you can desolder your joint (by reheating the solder), and reposition the components.

SOLDERING A CIRCUIT BOARD

STEP 1 Place the component that you wish to solder on the circuit board and clamp it down, then push its lead through one of the holes on the board.

STEP 2 Solder the leads to the bottom of the board. Press the soldering iron to the lead and the metal contact on the board at the point where you want them to connect. Once they heat up enough to melt the solder—just a few seconds—melt a drop of solder at the connection point.

STEP 3 Pull the solder away, then remove the soldering iron a second or two later. Once you've soldered all the leads onto the circuit board, trim off excess wire.

TRY TINNING

If you're working with components that have to be surface-mounted on a circuit board—ones that don't have leads you can thread through to the back of the board—use a technique called *tinning*.

STEP 1 Melt a drop of solder on the board where you want to attach the component. Then remove the soldering iron.

STEP 2 Pick up the component with tweezers, heat up the drop of solder on the board, and carefully place the component on the solder.

STEP 3 Hold the component in place for a few seconds until the solder cools.

STEP 4 If you need to desolder joints on a circuit board, use a desoldering pump.

003 STUDY CIRCUIT COMPONENTS

To build a circuit, you've got to know its building blocks.

Maps of how current flows through a circuit are called *schematics*. Symbols represent components, and lines show the current's path.

In this book, we use circuitry diagrams to show how to attach projects' components, so here we'll introduce you to some of the components that show up on these diagrams.

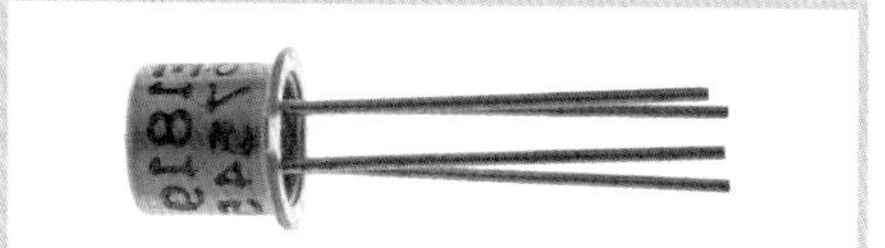

TRANSISTORS A transistor amplifies energy flowing to its base pin, allowing a larger electrical current to flow between its collector and emitter pins. The two basic types of transistors, NPN and PNP, have opposite polarities: Current flows from collector to emitter in NPN transistors, and flows from emitter to collector in PNP transistors.

POTENTIOMETERS When you need to vary resistance within a circuit, use a potentiometer instead of a standard resistor. These have a controller that allows you to change the level of resistance: "B" potentiometers have a linear response curve, while "A" potentiometers have a logarithmic response curve.

SWITCHES Switches open or close a circuit. Some are normally open as a default; others are normally closed.

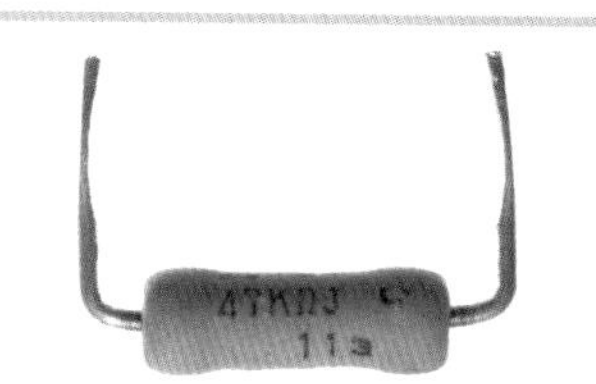

RESISTORS A circuit needs resistance to function. Without it, you'll end up with a short circuit, in which the current flows directly from power to ground without being used, causing your circuit to overheat and otherwise misbehave. To prevent that from happening, resistors reduce the flow of electrical current. The level of electrical resistance between two points is measured in ohms. Check to make sure a component's resistance matches the level indicated in the circuitry diagram.

BATTERIES These store power for a circuit, and you can use more than one to increase voltage or current.

CAPACITORS These store electricity, then release it back into the circuit when there's a drop in power. Capacitor values are measured in farads: picofarads (pF), nanofarads (nF), and microfarads (μF) are the most common units of measure. Ceramic capacitors aren't polarized, so they can be inserted into a circuit in any direction, but electrolytic capacitors are polarized and need to be inserted in a specific orientation.

WIRE These single strands of metal are often used to connect the components in a circuit. Wire comes in various sizes (or gauges), and it's usually insulated.

DIODES These components are polarized to allow current to flow through them in only one direction—very useful if you need to stop the current in your circuit from flowing the wrong way. The side of a diode that connects to ground is called the *cathode*, and the side that connects to power is called the *anode*. Light-emitting diodes, or LEDs, light up when current flows through them.

INTEGRATED CIRCUITS These are tiny circuits (usually including transistors, diodes, and resistors) prepacked into a chip. Each leg of the chip will connect to a point in your larger circuit. These vary widely in their composition, and will come with a handy data sheet explaining their functions.

TRANSFORMERS These devices range from thumbnail-size to house-size, and consist of coils of wire wound around a core, often a magnet. Made to transfer alternating current from one circuit to another, they can step the power of the current up or down depending on the ratio of wire windings between one coil and another.

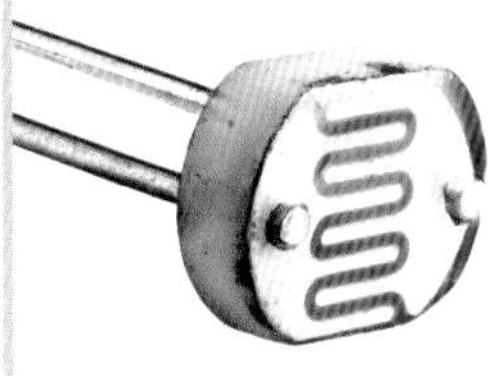

PHOTOCELL A photocell is a sensor that determines how much light (or other electromagnetic radiation) is present—it then varies its resistance between its two terminals based on the amount it detects.

004 BUILD A CIRCUIT

Now you know what goes into a circuit—so make one.

STEP 1 Assemble all the components that appear on your schematic, along with any tools you'll need.

STEP 2 To test your circuit before you solder it together, set it up on a breadboard. Breadboards are boards covered in small holes that allow you to connect components without soldering.

STEP 3 Once you're ready to construct the circuit, start by installing the shortest components. This helps you avoid having to move taller components out of the way. Orient labels in the same direction so they're all legible at once.

STEP 4 Many components have lead wires that you can insert into a circuit board. Bend these leads before you insert the component.

STEP 5 You'll need to hold your parts in place while you solder the circuit together. You can do this by clinching lead wires using tape, or bracing the parts against your work surface.

STEP 6 As you solder, check that each component is aligned correctly after you solder the first pin or lead—it's easier to make adjustments at this point, before you've finished soldering a part in place.

STEP 7 When everything's soldered in place, trim all your circuit's lead wires and test it out.

005 CHOOSE A MICROCONTROLLER

To be a geek, you've got to have these microcontroller basics under your belt.

A microcontroller is essentially a tiny computer with a central processing unit (CPU), memory, and input and output. It's useful for controlling switches, LEDs, and other simple devices. Scope these features.

PROGRAMMABILITY Some can only be programmed once, while some can be erased and reprogrammed. Some allow you to add external memory.

MEMORY Microcontrollers come with a set amount of memory. Make sure that the microcontroller you choose has sufficient memory to handle your project.

COMPLEXITY For a more complex project, seek out a model with lots of input and output pins and more memory than the lower-end microcontrollers.

PHYSICAL PACKAGING A microcontroller's construction can influence how easy it is to use. For instance, less space between pins can make the device harder to work with.

PROGRAMMING LANGUAGE Different microcontrollers use different programming languages. Choose one that uses a language you already know or are planning to learn.

SOFTWARE Some microcontrollers have easier-to-use software tools than others. If you're a beginner, ask around among your tech-savvy friends to get a sense of what's right for you.

006 PROGRAM AN ARDUINO

An Arduino is a an open-source microcontroller. Learn to program one and explore the possibilities.

STEP 1 Arduino microcontrollers come in a variety of types. The most common is the Arduino UNO, but there are specialized variations. Before you begin building, do a little research to figure out which version will be the most appropriate for your project.

STEP 2 To begin, you'll need to install the Arduino Programmer, aka the integrated development environment (IDE).

STEP 3 Connect your Arduino to the USB port of your computer. This may require a specific USB cable. Every Arduino has a different virtual serial-port address, so you'll need to reconfigure the port if you're using different Arduinos.

STEP 4 Set the board type and the serial port in the Arduino Programmer.

STEP 5 Test the microcontroller by using one of the preloaded programs, called *sketches,* in the Arduino Programmer. Open one of the example sketches, and press the upload button to load it. The Arduino should begin responding to the program: If you've set it to blink an LED light, for example, the light should now start blinking.

STEP 6 To upload new code to the Arduino, either you'll need to have access to code you can paste into the programmer, or you'll have to write it yourself, using the Arduino programming language to create your own sketch. An Arduino sketch usually has five parts: a header describing the sketch and its author; a section defining variables; a setup routine that sets the initial conditions of variables and runs preliminary code; a loop routine, which is where you add the main code that will execute repeatedly until you stop running the sketch; and a section where you can list other functions that activate during the setup and loop routines. All sketches must include the setup and loop routines.

STEP 7 Once you've uploaded the new sketch to your Arduino, disconnect it from your computer and integrate it into your project as directed.

THE FUN STUFF

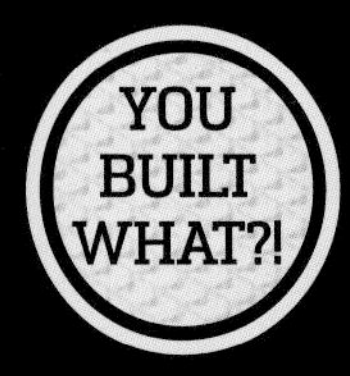

007 THE ULTIMATE ALL-IN-ONE BEER-BREWING MACHINE

Behold this deluxe homemade microbrewery: an elaborate device that boils, ferments, chills, and pours home-crafted ale.

What if there were a machine—a beautiful, shiny machine—and all it did, with almost no work from you, was make you beer? Such was the dream that drove former *PopSci* photographer John Carnett to build what he calls "the Device": a stainless-steel, two-cart brewing system that starts by boiling extract (concentrated wort, or pre-fermented beer) and ends with a chilled pint.

In most home-brewing setups, each step in the process requires moving the beer to a new container by hand, which increases the chance of contamination and requires lifting. Carnett's machine keeps everything in the carts' closed system—he only has to swap a few CO_2-pressurized hoses to move the liquid along.

The delicious brew's journey begins in the boil keg, where concentrated wort extract is heated by a propane burner for 90 minutes. The beer then travels through a heat exchanger—which cools the mix to about 55°F (13°C)—on its way to the fermenting keg. Here, a network of Freon-chilled copper tubes pumps cool water around the keg when the temperature gets too high. After two weeks, the Device pumps the beer into a settling keg, where a CO_2 tank adds carbonation. When you pull the tap, the beer travels through the cold plate, so it's chilled on the way to your glass. That's right: The Device is always ready with a cold pour and consumes no power when it's not actively serving or fermenting.

BUILDING A BETTER BREW The next step? Adding a third cart to make wort from raw grain instead of extract. But, says Carnett, there's a lot of "testing" of the new design to be done first.

008 DRINK BOOZE FROM A MELON

Turn basic produce into a hilarious drink dispenser.

MATERIALS

Medium seedless watermelon
Large spoon
Drill
Knife
Ball valve faucet with a handle
Rubber O-ring that fits the faucet
PVC-to-faucet adapter
Alcohol of your choosing

STEP 1 Using a knife, cut off just enough of the bottom of your melon so it sits flat.

STEP 2 Pick the side of the melon that you want to be the front, then cut a hole in the top, toward the rear. Save the piece you've cut out, as you'll use it later.

STEP 3 Scoop the fruit out of the melon with a large spoon.

STEP 4 Drill a hole in the melon's front, near the bottom. Using a knife, widen it so it's big enough for the faucet to fit inside.

STEP 5 Gently screw the faucet into the hole. (It helps to stick your free hand inside the melon and guide the faucet into place from the inside.)

STEP 6 Slide the O-ring onto the back of the faucet inside the melon, then install the adapter. Test for leaks.

STEP 7 Load it up with the elixir of your choosing, put the cut-out top back in place, and get your pour on.

009 BREAK INTO YOUR BEER

STEP 1 Use a metal file to wear down a carabiner's hook end so that it fits under a bottle cap's lip. (Be careful not to file it down too much or the carabiner won't close properly.)

STEP 2 Open the carabiner and place the unmodified end against the beer cap, then tuck the hook end under the cap's lip and use it as a lever to pry open your brew.

STEP 3 Carry your carabiner as a keychain so that you're always ready when beer suddenly, magically happens to you.

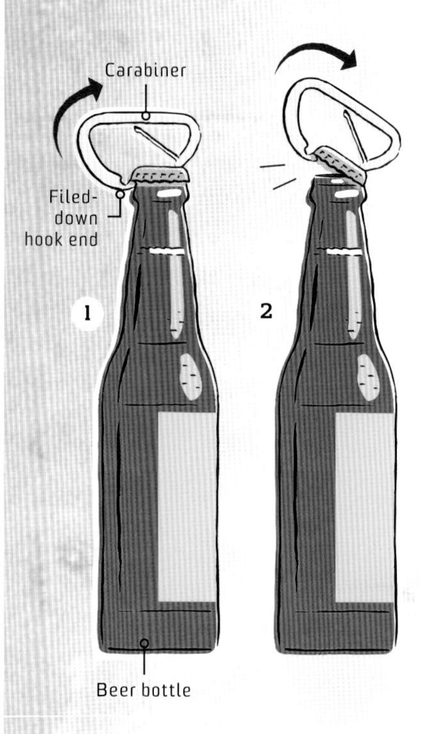

010 INSTALL A SHOWER BEER CADDY

STEP 1 Buy a cup holder at an automotive parts store. (Some have a hook on the back, which you don't need—remove it by gently breaking it off along the seam or cutting it off with a rotary tool.)

STEP 2 Drill a hole into the back of the cup holder that's just wide enough to accommodate the tip of a suction cup.

STEP 3 Insert the suction cup's tip into the hole, press the suction cup to the shower wall, and load your beer of choice into the caddy.

Beer can

Suction cup

Cup holder

011 CHILL YOUR BEER REALLY, REALLY FAST

STEP 1 Drill a hole into the side of a plastic container. The hole should be just wide enough that you can poke the straw of a container of compressed air through it.

STEP 2 Fill the plastic container with as many beers as you can fit. Cover with the lid.

STEP 3 Tape it shut. (Trust us. Otherwise the injected blast of cool air could blow the lid right off.)

STEP 4 Wearing thick, insulated gloves, turn the can of compressed air upside down and insert the can's straw into the hole, being careful not to touch the cans with it.

STEP 5 Squeeze for up to 1 minute.

STEP 6 Open the container and tap on each can's top for a few seconds to relieve the pressure inside. Then open one up and take a big swig—you've earned it.

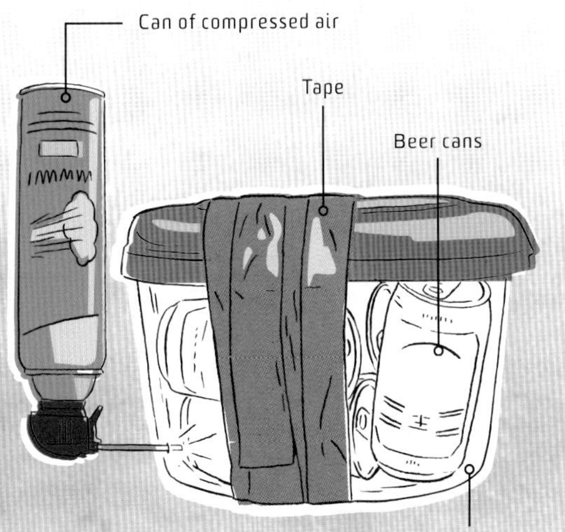

012 DISGUISE YOUR BREW

STEP 1 Using a can opener, remove the top and bottom of an innocent-looking soda can.

STEP 2 Using scissors, cut off the soda can's bottom lip.

STEP 3 Cut along the soda can's seam.

STEP 4 Using a metal file, sand down the can's edges to avoid cuts and scrapes. Get it as smooth as possible.

STEP 5 Wrap the soda can around your beer and enjoy your incognito beverage.

013 REUSE THOSE RED PARTY CUPS

Keep the infamous cup in play long after the keg's run dry. (Just rinse out that stale-beer smell first.)

DIY CAMERA LENS HOOD

Fend off glare with an impromptu lens hood. Cut the bottom out of your party cup, and then poke two holes on either side near the bottom. Thread a rubber band through each hole and knot them. Place the cup over your lens, tie the rubber bands around your camera near the viewfinder, and keep rain out of your shots.

MEASURE (MORE) BOOZE

Engineers say that the ridges on party cups are for structural integrity, but people have long used them as measurement lines: the first indentation from the bottom marks 1 ounce (30 ml), the second 5 ounces (147 ml), and the third a full 12 ounces (355 ml).

POOR MAN'S CHANDELIER

Poke holes in the bottoms of about 60 cups. Then glue the cups together along their sides with their open ends facing outward, forming a ball. Insert a bulb from a string of Christmas lights into each hole as you go to make a huge, sparkling plastic orb.

MAKESHIFT BEER INSULATOR

Cut Styrofoam to fit the bottom of your cup and place it inside, then spray a can with nonstick cooking spray. Put the can inside the cup and fill around it with some expanding foam. Let the foam dry and then trim the excess. Call it a koozie.

YOU BUILT WHAT?!

014 THE DRINK-SLINGING DROID

This robot tends bar like a pro. And even better, it never needs a tip.

A veteran of the TV show *BattleBots*, Jamie Price has built plenty of destructive machines. But recently he designed a robot with a more mellow calling: offering cold beer and cocktails. The result—a masterpiece of plywood, plastic, aluminum, and electric motors called Bar2D2—serves up everything but the sage advice.

The salesman modeled his machine on the iconic *Star Wars* droid R2-D2, and spent seven months and $2,000 building it. He used a plastic dome from a bird feeder as the head and built the robot's plywood skeleton to match. To make Bar2D2 mobile, Price stripped out the seat, the control system, and a pair of wheels from an electric wheelchair, added a new 12-volt battery, and wired a receiver to the motor so he could control it with an R/C helicopter-type remote.

Price fills each of the robot's six bottles with either liquor or a mixer, and then plugs these ingredients into a software program. The program computes a list of possible drinks, Price picks one, and the software sends pouring instructions to the robot via Bluetooth. A custom circuit board receives the signals and moves actuators that open specific valves just long enough for the robot's air-pressure system to force the right amount of each liquid into a waiting glass.

Bar2D2 has already proven to be a hit among robotics and cocktail fans alike, but Price isn't finished yet. Next he's adding a breathalyzer and an LED-backed projector that displays blood-alcohol content. Give us your keys, Obi-Wan.

BEER ME, BAR2D2

One difficulty was finding a way to move bottles up from the enclosed beer rack to the serving station above. When Price hits a button on his remote, the rod of a motorized caulk gun extends and pushes the beer up from the lower level. He calls it his beer elevator.

5 MINUTE PROJECT

015 SERVE SHOTS IN JELL-O CUPS

STEP 1 Mix up your Jell-O and pour it into a muffin tin's compartments, filling each about half full.

STEP 2 Place small pebbles into paper cups to weight them. Put a cup in each tin compartment and cover each with tape.

STEP 3 Wait for the Jell-O to solidify around the paper cups, molding into a cup shape. Remove the cups and pebbles.

STEP 4 Fill up the Jell-O cups with booze. Be popular at a party near you.

016 MAKE DRINKS GLOW IN THE DARK

The magic glowing ingredients? Simple riboflavin and quinine, plus a trippy black light.

FOR BLUE DRINKS:

STEP 1 Mix any drink you want.

STEP 2 Add tonic water.

STEP 3 Drink it near a black light.

FOR YELLOW DRINKS:

STEP 1 Crush up a B2 vitamin and put a pinch of the powder in the bottom of your glass.

STEP 2 Pour in a flavored drink, as the vitamin has a faint bitter taste.

STEP 3 Drink it near a black light.

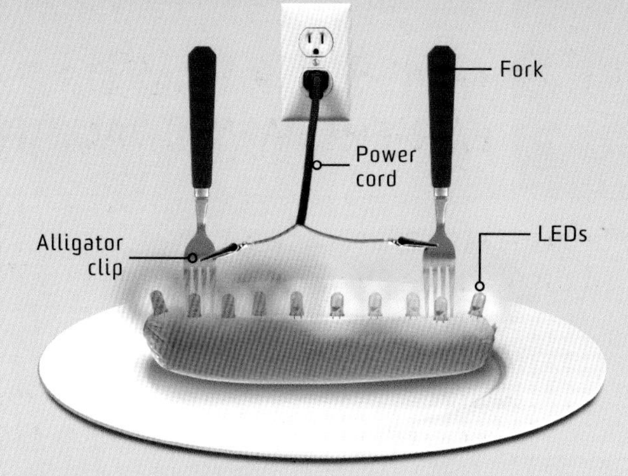

017 COOK A HOT DOG WITH ELECTRICAL CURRENT

An LED display lights up the room and nukes your hot dog, too.

MATERIALS

Power cord
Wire strippers
Two alligator clips
Soldering iron and solder
Hot dog
Nonconductive plate
Two metal forks
Assorted LEDs

STEP 1 Using wire strippers, cut off the end of the power cord and peel back the outer insulation. Next, snip back the green ground cord and then strip the ends of the remaining two wires.

STEP 2 Solder one small alligator clip to each of the stripped wires (except for the ground wire).

STEP 3 Put the hot dog on a nonconductive plate (ceramic works nicely). Secure each alligator clip to a fork, and stick the forks into the hot dog.

STEP 4 Stick some LEDs into the hot dog.

STEP 5 Very carefully plug the cord into the wall. Don't touch the hot dog or any of the rest of the contraption while the cord is plugged in.

STEP 6 The hot dog will cook in a minute or two. Not that you're going to eat it, right?

WARNING
Use this activity to impress your friends with your electrical chops, not your culinary skills. Eating the resulting hot dog is a seriously bad idea. Also, keep water very far away from this science experiment!

018 MOD YOUR TOASTER FOR FAR-FROM-AVERAGE TOAST

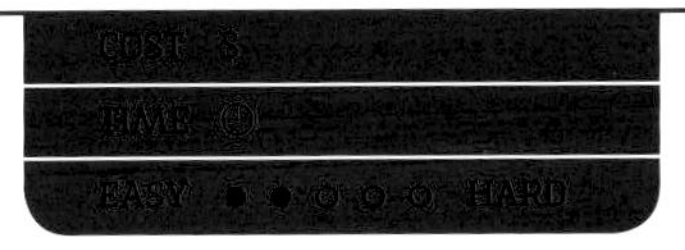

Because bread tastes a lot better with funny faces on it.

MATERIALS

- Toaster
- Paper
- Pencil
- Craft knife
- Glue stick
- Aluminum flashing
- Utility knife
- Metal file
- Bread that needs toasting

STEP 1 Unplug your toaster and remove the insert—that's the part that holds bread slices in place. Measure the space between the insert's two prongs.

STEP 2 Cut two pieces of paper to fit between these two prongs. Draw or print a shape that you want to see on your breakfast onto the paper. Include a tab on either side of each shape that you can wrap around the prongs to hold your design in place inside the toaster.

STEP 3 Use a craft knife to cut out the negative spaces around and in your design, creating a stencil.

STEP 4 Glue the paper pieces to aluminum flashing. Use a utility knife to cut the shapes and their tabs out of the flashing, then smooth the edges with a metal file.

STEP 5 Gently and thoroughly wash off the glue and remove all the paper. With the toaster unplugged, use the shapes' tabs to hook them to the toaster insert.

STEP 6 Plug the toaster in, replace the insert, put in some bread, and make some really fun toast.

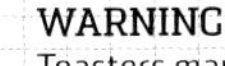

WARNING
Toasters may seem harmless enough, but once they're plugged in they're juiced with powerful voltage. No prongs, aluminum flashing, or paper bits should come into contact with the toaster's electric heaters.

019 FREEZE LEGO ICE CUBES

Make a mold of LEGO bricks and enjoy the world's geekiest ice.

MATERIALS

A LEGO base plate and bricks
Petroleum jelly
Mold compound
Craft knife
Food coloring, if desired

STEP 1 Wash and dry the base plate and blocks.

STEP 2 Build a LEGO tray on the base plate: Make four sides three blocks in height, then place single bricks inside the border, spacing them evenly with one or two rows of bumps between each.

STEP 3 Coat the tray in petroleum jelly, then slowly fill the tray with the mold compound and set aside for at least 12 hours. Bang on the table to prevent bubbles.

STEP 4 Peel the mold out of the LEGO tray and trim away any random, clinging bits with the craft knife.

STEP 5 Wash the mold and turn it over. Fill the depressions with water (add food coloring, if you roll that way), slide it in the freezer, and await your cubes.

020 THE PROPANE-POWERED FIREGUN

Why go to a lot of trouble just to shoot fireballs over the Philadelphia skyline? The better question is, why not?

Fire enthusiasts have long used propane "poofers" to shoot huge fireballs for special effects. But for this particular model, *PopSci* contributor Vin Marshall tried a new approach that incorporates striking visual elements as well as a bit of science. This model has a cone-shaped mixing burner known as an inspirator, which uses the pressure of the escaping propane to form a vacuum that draws in atmospheric oxygen.

This mixture creates more-complete—and louder—combustion, producing a sound like a rocket launch and a ferocious column of flame reaching more than 20 feet (6.1 m) in the air. Combining that burner with a valve that opens instantly when electrical power is applied allows Marshall to send up one intense flame after another with the push of a button. A Buck-Rogers–style gas tank adds effect.

It took 40 crazed hours of near-nonstop parts acquisition, construction, and testing in a friend's partially collapsed warehouse to finish the poofer. No doubt it was all worth it.

SECOND LIFE
Marshall repurposed parts to build his contraption on a budget. The switch box, which controls power to the solenoid valve, was once used to start and stop industrial machines.

021 MAKE A SONIC TUNNEL OF FIRE

See your favorite song burst into flame with this classic Rubens' tube.

COST $$

TIME ◷◷◷

EASY ● ● ● ○ ○ HARD

This may be one of the best bad ideas of all time, and we have physicist Heinrich Rubens to thank for it: He found that if you make a sound at one end of a tube, you get a standing wave equivalent to the sound's wavelength inside the tube—and that the best way to demonstrate this principle is with waves of flame synced with music. Right on, Heinrich.

MATERIALS

- 4-inch (10-cm) ventilation ducting
- Nail
- Drill
- Duct tape
- Latex sheets
- Scissors
- Two hose splicers
- Epoxy putty
- Hose T-connector
- Propane tank
- Teflon tape
- Two 4-inch (10-cm) brackets
- Screws for your brackets
- Scrap wood
- Media player and speakers

STEP 1 Leaving 4 inches (10 cm) at either end, mark off every 1/2 inch (1.25 cm) down the length of your ducting. (Do it on the side without the seam.)

STEP 2 Gently tap a nail at each interval, creating divots that will be easy to drill. Then drill through each depression.

STEP 3 Wrap a strip of duct tape around each end of the tube. Then cut two squares of latex and tape them across both open ends of the tube, creating an airtight seal.

STEP 4 Select two spots for fuel entry in the seam, each about one-third of the way across the tube. Tap the locations with a nail to create depressions, then drill two holes large enough for your hose splicers.

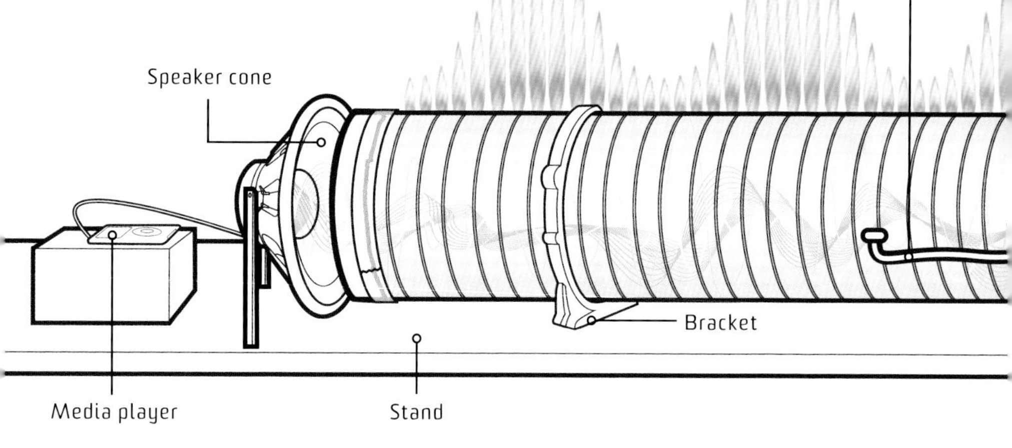

WARNING

Things that shoot flames are highly dangerous—even more so when set next to a full jug of propane. Exercise extreme caution while building and playing your Rubens' tube, and always have a fire extinguisher at the ready.

STEP 5 Install the hose splicers, securing the edges around the fuel entry holes with epoxy putty.

STEP 6 Attach the T-connector to the propane nozzle, then the hose splicers to the T-connector. Wrap the ends of all the components with Teflon tape.

STEP 7 Using screws, attach brackets to scrap wood to make a stand. Then mount the tube onto it.

STEP 8 To use, tape all holes to create a seal. Then pump the tube full of propane for 2 minutes.

STEP 9 Remove the tape and test the tube by lighting one hole. If the flames are 1 inch (2.5 cm) high, it's ready.

STEP 10 Place a speaker as close as possible to one end of the tube (without actually touching the end's latex seal). Hit play, and watch those sound waves ignite.

THE RUBENS' TUBE IN ACTION

Why just listen to "Light My Fire" when you can listen to it *and* see its sound wave expressed in real flames?

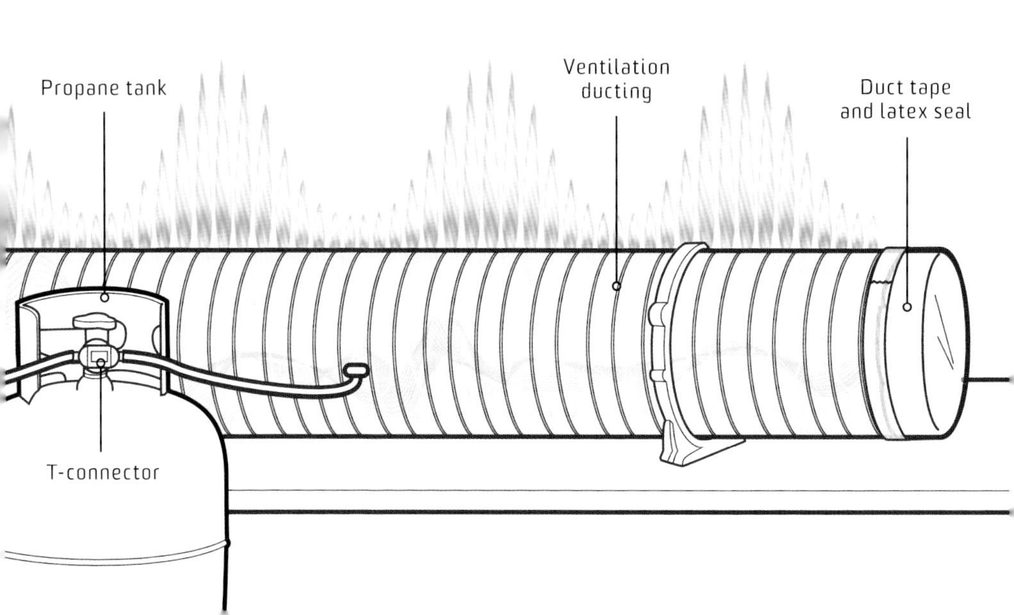

022 TURN YOUR CAMPFIRE GREEN

STEP 1 Pour 1/4 inch (6.35 mm) of copper sulfate into a small paper cup. (You can use common tree-root killers, which contain copper sulfate.)

STEP 2 Melt old candle stubs in a double boiler, and pour the wax into the cup over the copper sulfate.

STEP 3 Stir the copper sulfate and wax together until the chemical is coated.

STEP 4 After it cools, peel off the paper cup.

STEP 5 When you're done cooking at your campsite, throw the copper-sulfate-infused wax into the hottest part of the fire and watch the green flames start licking.

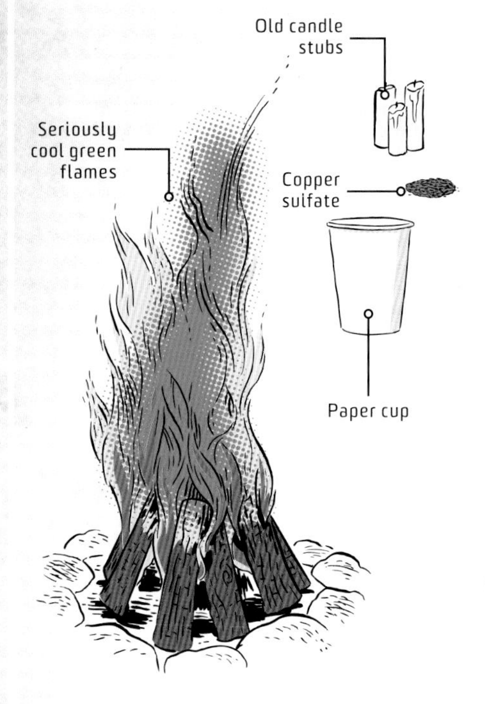

023 HOLD A FLAMING BALL IN YOUR BARE HAND

STEP 1 Use scissors to cut away a 2-by-5-inch (5-by-12.5-cm) strip of cloth from an old T-shirt. Roll the cloth into a ball.

STEP 2 Thread a needle with about 2 feet (60 cm) of sewing thread.

STEP 3 Push the needle all the way through the fabric ball, securing the loose end of the fabric strip.

STEP 4 Wind the thread around the ball many, many times. When you're almost out of thread, pull the needle through an existing loop of thread, then tie it off and remove the needle.

STEP 5 Soak the ball in isopropyl alcohol; squeeze out any excess that may drip onto your hands.

STEP 6 Wash any fluid off your hands, light up your fireball, and let it blaze around in your hand. (The less-adventurous can put on heat-resistant gloves.)

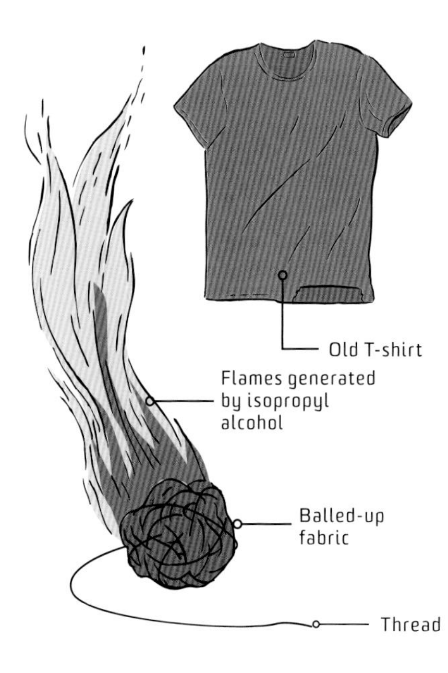

024 SET OFF A SPINNING FIRE TORNADO

STEP 1 In the center of a lazy Susan, mold clay into a base for a fireproof bowl. Press the bowl into the clay, and place pieces of modeling clay along the edges of the lazy Susan.

STEP 2 Measure the lazy Susan's diameter and roll a piece of screen into a 36-inch- (90-cm-) high cylinder of the same diameter. Use straight pins to secure the cylindrical shape.

STEP 3 Pour kerosene onto a rag; place it in the bowl.

STEP 4 With a fire extinguisher nearby, carefully ignite the rag in the small bowl with a long-handled lighter, then place the screen cylinder over the lazy Susan, pressing it into the pieces of molding clay.

STEP 5 Give the lazy Susan a whirl. Stand back and watch devious, fiery nature at work.

STEP 6 To extinguish, don heat-resistant gloves, wait for the lazy Susan to slow, and remove the screen. Then snuff out the small bowl with a larger fireproof bowl.

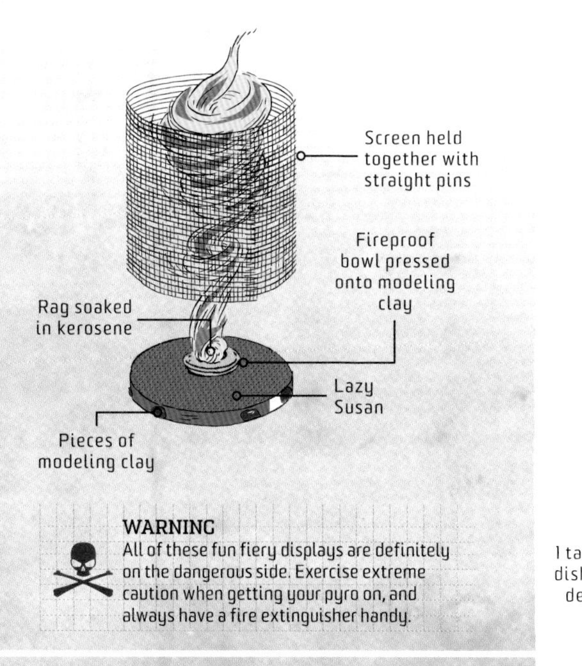

WARNING
All of these fun fiery displays are definitely on the dangerous side. Exercise extreme caution when getting your pyro on, and always have a fire extinguisher handy.

025 IGNITE A HOMEMADE SPARKLER

STEP 1 Using a rotary tool, make a small hole into the top of a medicine bottle's lid.

STEP 2 Fill the bottle about one-fourth of the way with water and add 1 teaspoon of salt.

STEP 3 Add 1 tablespoon each of powdered dishwashing detergent and baking soda, and then put the lid on.

STEP 4 Carefully hold the flame of a lighter or match over the hole until the gas you've just created ignites, firing your sparkler. It's time to celebrate.

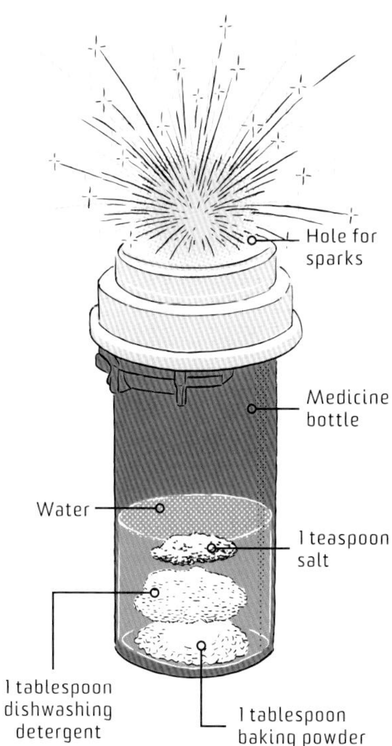

026 WREAK HAVOC WITH THE ULTIMATE SQUIRT GUN

COST	$$
TIME	◷ ◷
EASY	● ● ○ ○ ○ HARD

Water guns have never been more fun. No, really, we mean it.

This wet weapon is way more than a squirt gun; it's a powerful water cannon that shoots more than 1 quart (950 ml) of water up to 50 feet (15 m) in less than 10 seconds. But don't be a jerk. Keep your H2O gun's spray away from other people's faces.

MATERIALS

- Drill
- Various PVC parts (see diagram below)
- PVC cement
- Waterproof grease
- Bucket of water

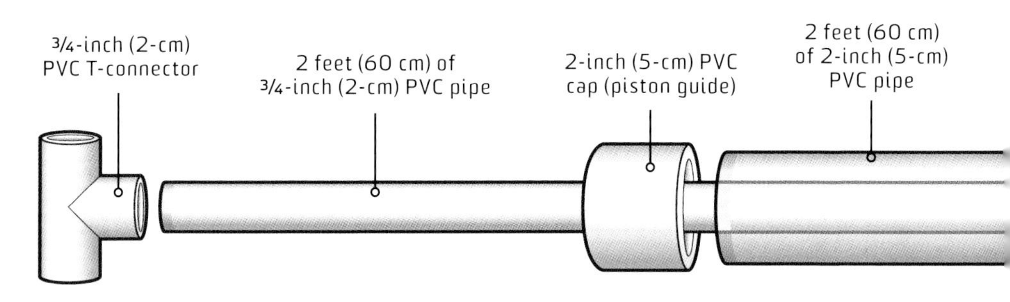

STEP 1 Use a drill to make a 1/4-inch (6.35-mm) hole in the center of one 2-inch (5-cm) cap and a 1 1/4-inch (3-cm) hole in the center of the second cap. The first is your nozzle; the second is the piston guide.

STEP 2 Glue the T-connector to the small pipe with PVC cement. When gluing, immediately insert the pipe into the fitting and turn it to distribute the solvent evenly. Hold the joint for about 30 seconds to make sure it sets; wipe off excess glue with a rag.

STEP 3 Slide the piston guide over the small pipe with the open end facing away from the T-connector.

STEP 4 Glue the reducer bushing to the small pipe's end, the coupler to the reducer, and the 1 1/4-inch (3-cm) pipe to the coupler with the PVC cement.

STEP 5 Slide the O-ring over the small 1 1/4-inch (3-cm) pipe and glue the 1 1/4-inch (3-cm) PVC cap to the pipe.

STEP 6 Glue the nozzle onto the big pipe. Let the apparatus dry.

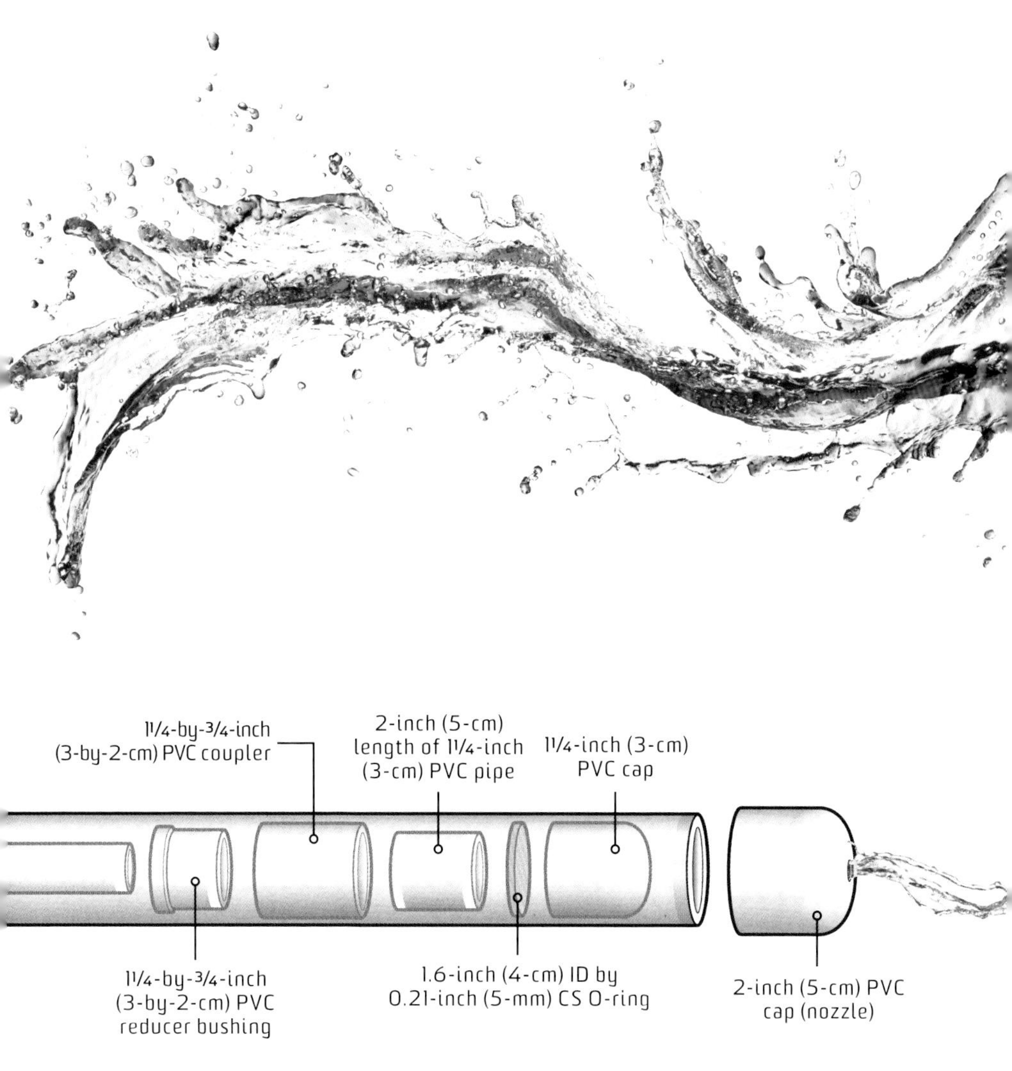

STEP 7 Apply a small glob of waterproof grease to the inside of the 2-inch (5-cm) PVC pipe. Insert the piston into the body and push and pull a few times to evenly spread the goop. When it seems sufficiently lubricated, firmly push the piston guide onto the body.

STEP 8 To load, use it like a giant syringe: Compress the handle and stick the huge squirt gun's end into a bucket of water, then pull up on the T-connector to draw water into the pipe.

STEP 9 Super-soak somebody near you.

WARNING
This thing can really let loose, so be cautious about spraying it at living creatures. (Zombies, though, you can totally let have it.)

027 BUILD A BRISTLEBOT

All the fun of a hyperactive pet, minus all the annoying shedding.

MATERIALS

Toothbrush with angled bristles
Rotary tool
Double-sided foam tape
Pager motor
Glue
Coin battery
Electrical tape
Decorations

STEP 1 Use a rotary tool to cut the head off a toothbrush. Apply double-sided foam tape to the back.

STEP 2 Salvage a pager motor with two wires and connect the wires to a coin battery, positive to positive and negative to negative. Tape the wires in place on the battery with electrical tape.

STEP 3 Add decorations, then attach the motor and battery to the foam tape on the toothbrush. Watch the robot merrily frolic.

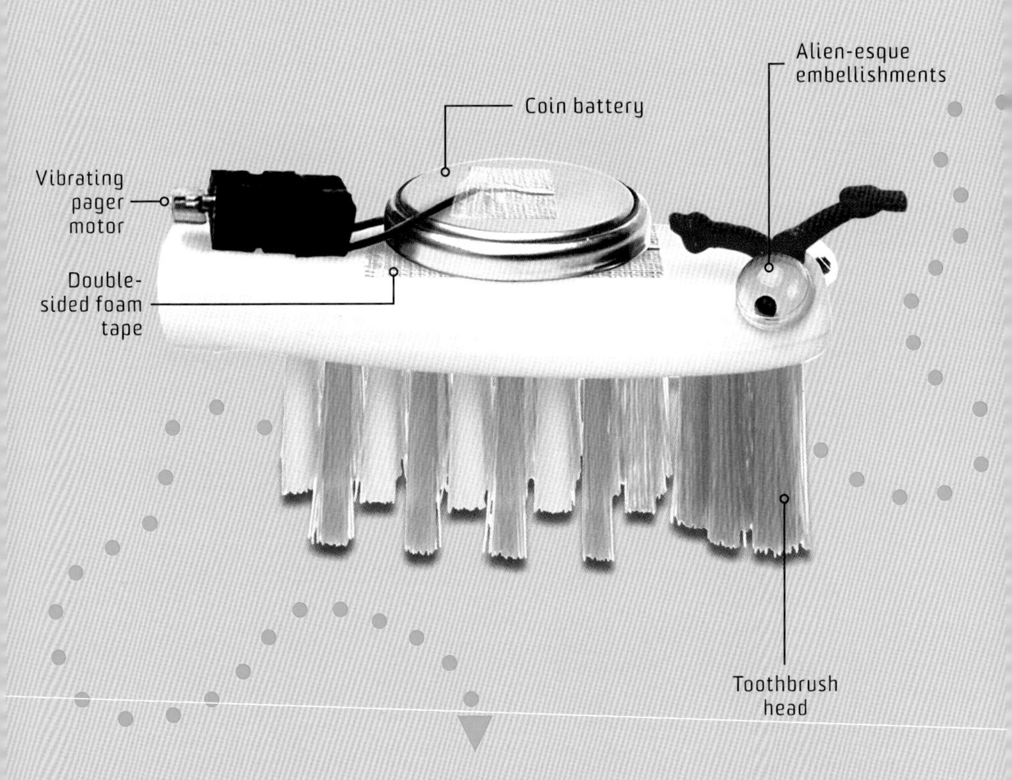

028 Amuse Yourself with a Flipperdinger

Make uncles everywhere jealous with this dorky device, which makes a ball hover in midair.

MATERIALS

- Putty
- Long hollow reed
- Knife
- Smaller hollow reed
- Acorn cap
- Glue
- Small, lightweight ball

STEP 1 Put putty into one end of the long reed, making an airtight seal. Near this end, make a hole through the reed. Stick the smaller reed into this hole, making a nozzle.

STEP 2 Remove the cup-shaped cap of an acorn and poke a hole in its center. Then fit the cup over the nozzle and secure it with glue.

STEP 3 Place the ball in the acorn cap and blow lightly but steadily into the open end of the reed. When done right, the ball rises slowly on a jet of air, hovering above the nozzle. As you ease off, the ball settles back into the cap, to the wonderment of all.

029 MAKE A MINI CATAPULT

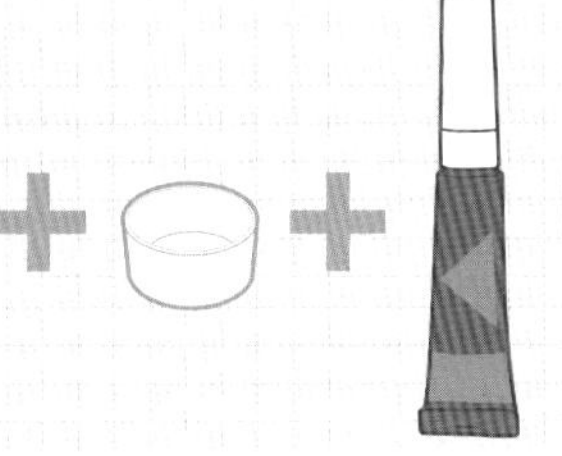

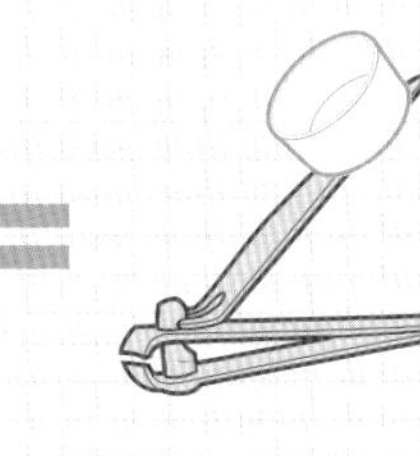

STEP 1 Lift up and rotate the nail clipper's little lever.

STEP 2 Glue the bottle cap onto the lever, leaving a little room for your fingertip at the very end.

STEP 3 Load it up with supersmall ammo, pull it back, and let go to launch.

030 IMPROVISE A PLANETARIUM

COST	$$
TIME	◴◴◴
EASY	● ● ● ● ○ HARD

Gaze at a starfield featuring twinkling constellations of your own devising.

MATERIALS

- LilyPad Arduino
- Breakout board
- USB cable
- Velcro
- Sewing needle and thread
- Two same-size pieces of black fabric
- Conductive thread
- Wire strippers
- Single-stranded wire
- Six LEDs
- Soldering iron and solder
- Scissors
- Fiber optic filament
- Electrical tape
- 3.7-volt polymer lithium ion battery and a mini USB charger for them
- Small clear beads
- Hot-glue gun

STEP 1 Connect your LilyPad Arduino to your computer using a breakout board and a USB cable. Then load it up with the code at popsci.com/thebigbookofhacks.

STEP 2 Sew Velcro around the edges of the fabric pieces, and sew the LilyPad Arduino near the edge of one fabric piece using conductive thread.

STEP 3 Strip the wire and make small loops. Solder the loops to the six LEDs' connectors, making "buttons."

STEP 4 Create a pattern and print it out in a size to fill your ceiling. Tape it onto

the other piece of fabric—not the one you attached the LilyPad Arduino to.

STEP 5 Look at your pattern and decide where you'll need the most fiber-optic bundles. Space the LEDs so the filaments can extend from them to fill the pattern.

STEP 6 Sew the buttons onto the fabric piece that you sewed the LilyPad Arduino to, connecting each LED to one of the LilyPad Arduino terminals with conductive thread. Use terminals 3, 5, 6, 9, 10, and 11.

STEP 7 Cut fiber-optic filaments into varying lengths and gather them into six bundles of 10 to 20 strands. Then secure the ends of the filaments in each bundle together with tape.

STEP 8 Attach the battery to the LilyPad Arduino; each LED will light up. Then use electrical tape to secure the filament bundles over the LEDs and to the fabric.

STEP 9 Thread the filaments through the second piece of fabric, following the pattern. It helps to use a small, sharp tool to poke holes for the filaments. When finished, remove the pattern.

STEP 10 Slide a clear bead onto each filament and hot-glue it on the underside of the fabric. Trim the filament.

STEP 11 Use the Velcro to connect the fabric pieces together with the filaments in between them. Hang it up with small nails or hooks, lean back, and admire your new starry, starry night.

031 JAM OUT TO A SOUND-REACTIVE LIGHT BOX

Watch beats blink in time with a slick-looking LED display.

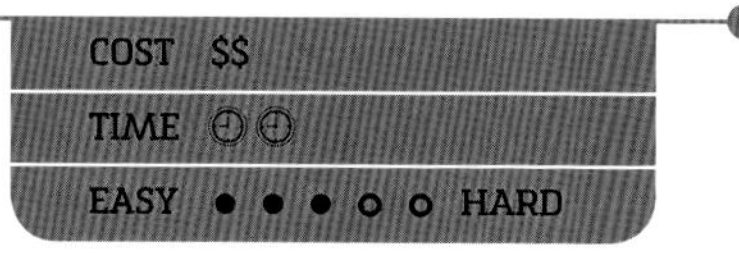

MATERIALS

- Sheet of 3-mm Plexiglas
- Ruler
- Table saw
- Drill with a glass bit
- Audio cable
- 18-volt adapter
- Fine sandpaper
- Hot-glue gun
- Six white 5-mm LEDs
- TIP31 transistor
- Electrical wire
- Soldering iron and solder

STEP 1 To make the box, measure the Plexiglas into four 6-by-2-inch (15-by-4.7-cm) pieces and two 2-by-2-inch (5-by-5-cm) pieces. Cut them with a table saw outfitted with a plastic-cutting blade.

STEP 2 Drill two holes near a corner in one of the long pieces: one for the audio cable that will go to a stereo and one that's large enough to fit the plug on the adapter cord. Go lightly or the Plexiglas may break.

STEP 3 Using a circular motion, sand both sides of the pieces and the surfaces of the LED bulbs until you've achieved a cloudy, frosted look.

STEP 4 Hot-glue three of the rectangular panels together along their long edges, then glue the square pieces to the ends. Sand the joints after the glue dries.

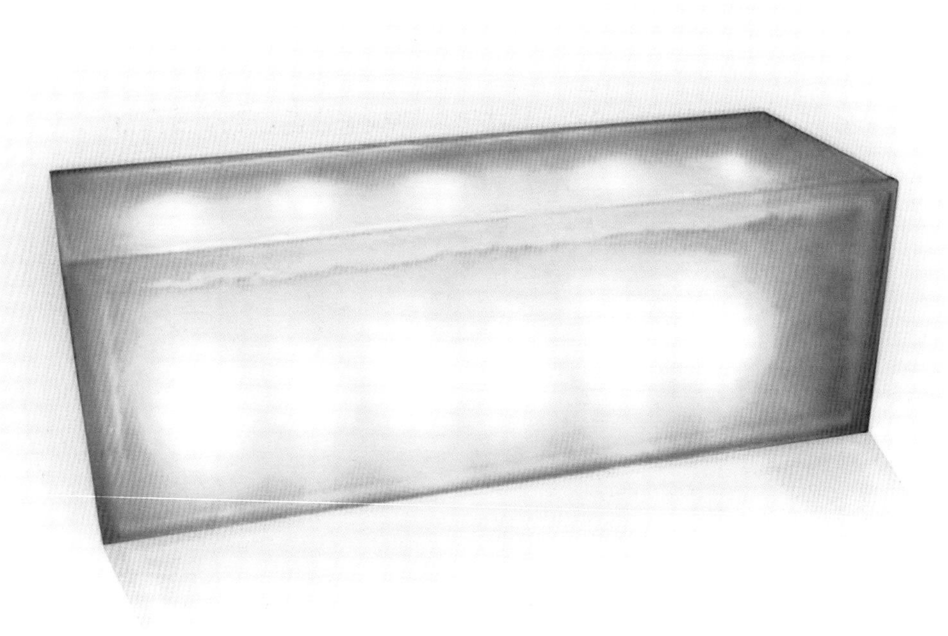

STEP 5 Pull the audio cable through one of the box's holes and peel back the plastic to expose its wires on the inside of the box.

STEP 6 Wire the electronics according to the circuitry diagram. If you want to add more LEDs, buy an adapter that provides each LED with 3 volts.

STEP 7 Put the circuit in the box. Pull the adapter plug through the hole in the box, then glue it in place.

STEP 8 Glue the last Plexiglas piece onto the box. Plug the audio cable into your stereo's speaker output and plug the adapter into a power outlet.

STEP 9 Pick a song, and see it in lights.

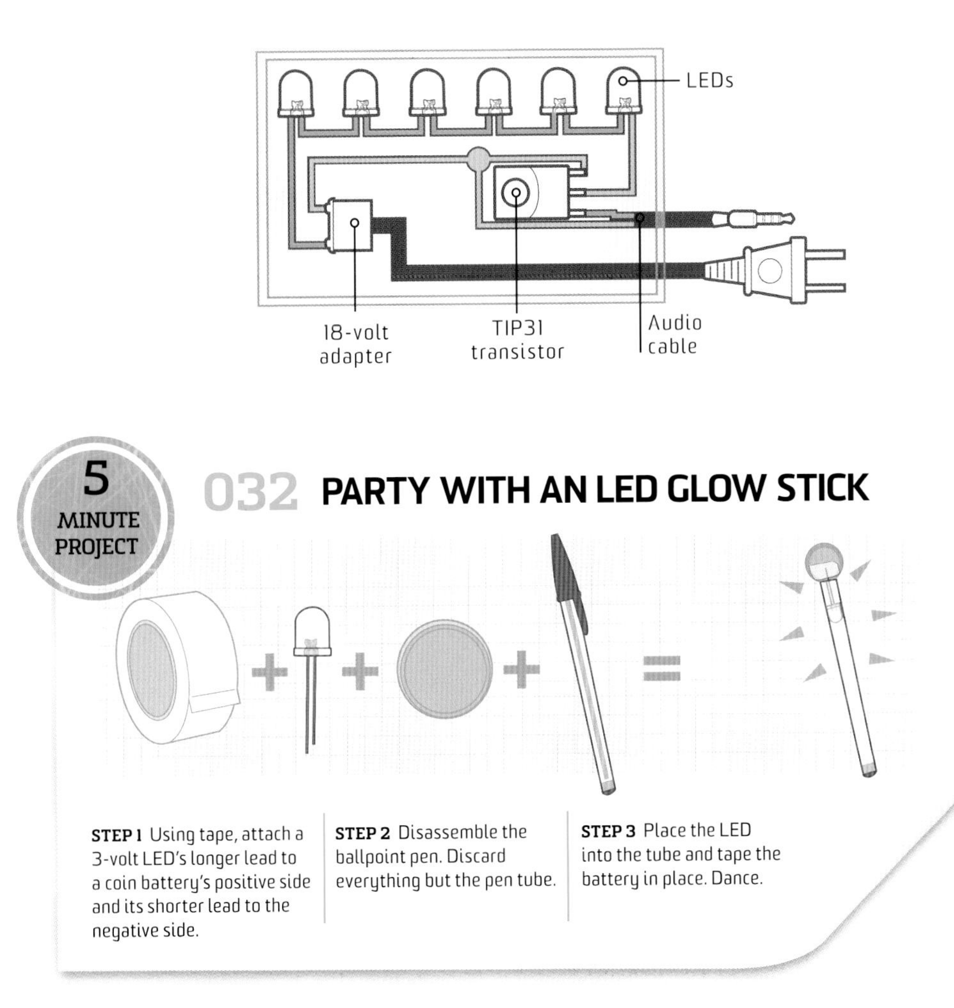

032 PARTY WITH AN LED GLOW STICK

STEP 1 Using tape, attach a 3-volt LED's longer lead to a coin battery's positive side and its shorter lead to the negative side.

STEP 2 Disassemble the ballpoint pen. Discard everything but the pen tube.

STEP 3 Place the LED into the tube and tape the battery in place. Dance.

033 WAVE AN LED LIGHTER AT A CONCERT

Power ballads sound even more epic with this lighter mod.

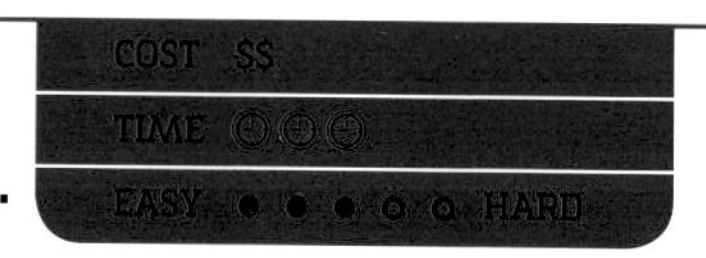

MATERIALS

- Dead lighter with absolutely no fluid inside
- Pliers
- Hacksaw
- Rotary tool
- 3-volt LED
- Soldering iron and solder
- Electrical wire
- Two AAA batteries
- Aluminum-foil duct tape
- Superglue

STEP 1 Check to make sure your lighter is empty. If not, hold down the lever until the lighter fluid evaporates.

STEP 2 Using the pliers, pry off the metal shield at the top of the dead lighter.

STEP 3 Carefully remove the striker wheel, fuel lever, spring, and fuel valve inside; set these parts aside.

STEP 4 Cut 1/4 inch (6.35 mm) off the bottom of the lighter with the hacksaw. Pry out the plastic divider—this may require you to use a rotary tool.

STEP 5 In the middle of the metal shield's underside, create a dent with your pliers to serve as a contact point for the switch.

STEP 6 Solder the LED's negative lead to the shield near the dent and the positive lead to a piece of electrical wire 1 inch (2.5 cm) in length.

STEP 7 Solder a piece of wire 1 inch (2.5 cm) in length to the underside of the metal part of the fuel lever.

STEP 8 Put the spring and fuel lever inside the lighter. Reinsert the metal shield and thread the long wire attached to the LED through the flint tube.

STEP 9 Line the batteries up with opposite polarities next to each other, then tape a piece of stripped wire across them using aluminum-foil duct tape.

STEP 10 Follow the circuitry diagram, connecting the LED and fuel lever's wires to the batteries using the aluminum-foil duct tape.

STEP 11 Slide the batteries inside and glue the bottom back on. It's officially slow-jam time.

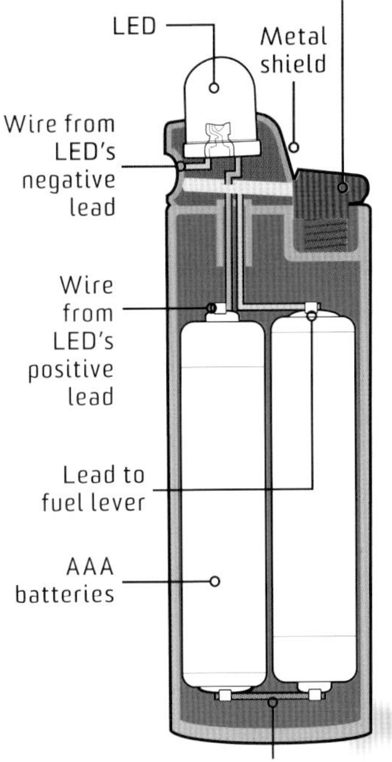

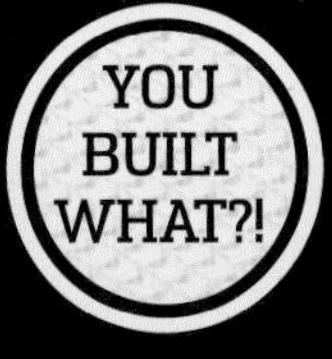

034 AN LED-LIT DISCO DANCE FLOOR

When dance fever hit MIT, students built a computer-controlled, LED-lit disco floor.

A group of undergrads at the Massachusetts Institute of Technology took on a challenge more daunting than classwork: disco. Before a dorm party, Mike Anderson, Grant Elliott, Schuyler Senft-Grupp, and Scott Torborg worked night and day for a week to build a computer-controlled, pixelated dance floor out of 1-by-4-foot (30-by-120-cm) boards, LEDs, tinfoil, paper towels, and old computer parts. The result would make Travolta weep with joy.

Each of the 512 pixels contains three LEDs pointed down at a square of paper towel that sits in a larger piece of foil. The foil reflects the light up through the plastic floor, while the paper towel mutes its glow. (LEDs stay cool, so the towels won't ignite.) A computer controls each pixel individually, and the open-source software generates 25 disco-tastic patterns, enabling DJs to match the light show to the music they're playing—and code-savvy disco fans to add new light patterns. What's more, by varying the intensity of each bulb, the students can blend light from the red, green, and blue LEDs housed in each pixel to produce any color. And should the party get extra wild (and with a dance floor like this, it will), the platform's wooden frame and thick layer of Lexan plastic make it nearly indestructible.

After earning minor fame at MIT (one of the inventors scored dates because of his uncanny soldering skills), the students began upgrading the floor. Their latest model has a prebuilt circuit board and instructions, so anyone can turn a basement into a discotheque.

HOW GEEKS GET DOWN
Everyone who actually worked on the floor and isn't a professional model, raise your hand.

035 PUMP JAMS THROUGH AN OLD-SCHOOL PHONOGRAPH

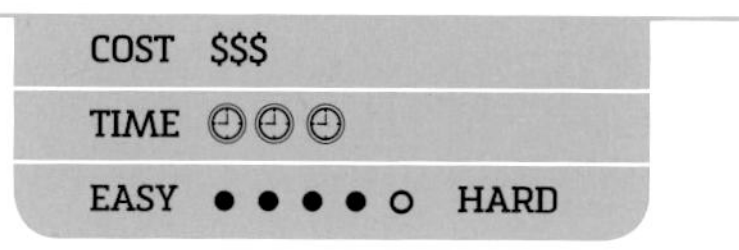

Give a phonograph new life with this modernizing mod.

MATERIALS

- Wooden box
- Drill with a hole bit
- Felt
- Miniature mono amplifier with tone control
- Transformer for the amplifier
- Two potentiometer knobs
- 3.5-mm stereo socket
- Power plug
- Power switch
- Speaker
- Soldering iron and solder
- Electrical wire
- Hot-glue gun
- Brass horn from an old phonograph
- Media player

STEP 1 Measure and cut two holes in the top of your box: one to fit the brass horn and a smaller one that will fit the stereo socket.

STEP 2 Measure the box's inside and cut felt to those dimensions, then line the box with felt—it will make for better sound.

STEP 3 Set up the electronics according to the circuitry diagram, drilling holes for the power source, potentiometers, and power switch as you go.

STEP 4 Place the horn in its hole and hot-glue it in place. Sand and varnish the box, if you desire.

STEP 5 Close up the box, plug your media device into the stereo socket, and plug the contraption's power cable into an outlet.

STEP 6 Enjoy the sweet, sweet sound of anachronism.

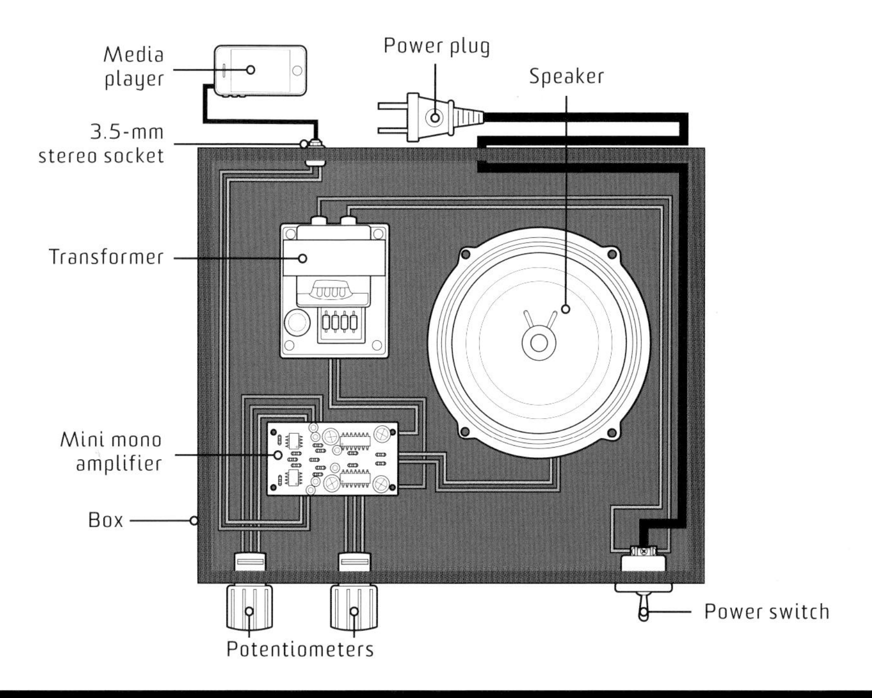

036 MAKE CUSTOM-FIT EARBUDS

Because anything you stick in your ear every day should be comfortable.

MATERIALS

Earbuds
Craft knife
Silicone putty

STEP 1 Use a craft knife to remove the stock tip (usually made of foam or rubber) that your buds came with.

STEP 2 Follow the instructions on your putty package to get the putty ready.

STEP 3 Pull up on the tip of an ear and, with your mouth open, press some putty in gently, folding in or removing the excess to create a flush, clean fit.

STEP 4 With the putty still in your ear, insert the bud.

STEP 5 After the silicone has set (about 10 minutes), remove the mold by gently twisting it out of your ear.

STEP 6 Gently pull the bud out of the mold and use the craft knife to make a small hole in the mold that will allow sound to come through. Reinsert the bud.

STEP 7 Repeat the entire process on the other earbud and enjoy a personalized fit.

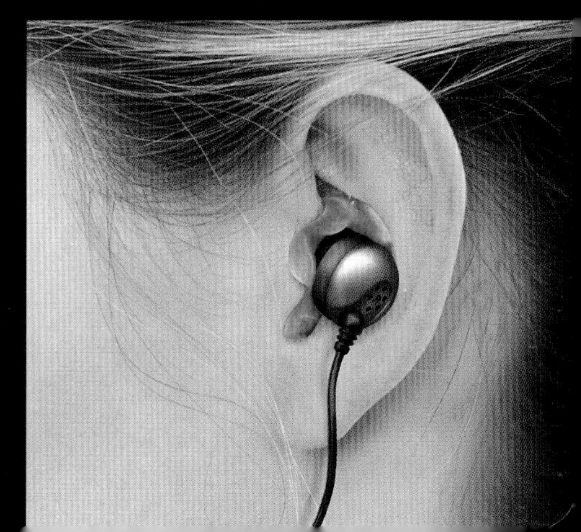

037 AIR DRUM WITH ELECTRIC DRUMSTICKS

COST	$$$$
TIME	◷◷◷◷
EASY	● ● ● ● ● HARD

Now the masses can finally hear your brilliant air drumming.

This very cool DIY drumming setup uses an Arduino and a MIDI—that's short for "musical instrument digital interface"—device to create a drum kit sound without the drum kit price tag. Move the drumsticks to hit the snare or cymbal; tap your feet to hear the bass or hi-hat.

MATERIALS

- Two 1½-inch (3.75-cm) wooden dowels, about 1 foot (30 cm) in length
- Drill
- Eight USB female type A connectors
- Three ADXL335 accelerometers
- Electrical wire
- Wire strippers
- Soldering iron and solder
- Two 1½-inch (3.75-cm) vinyl end caps
- Rubber-soled shoes you don't mind donating to the cause
- Photocell
- 47k-ohm resistor
- Four USB male-to-male type A cables
- Arduino UNO
- Five-pin DIN connector
- 5-volt DC-power supply with 2.1mm jack
- USB-to-MIDI interface

STEP 1 Drill a hole through the two dowels—these will be your drumsticks. In each drumstick, widen the holes to fit a USB female connector at one end and an accelerometer at the other end.

STEP 2 Follow the circuitry diagram to solder together your cymbal and snare, housing their circuitry in the left and right drumstricks, respectively. Cover the ends of the dowels with vinyl end caps.

STEP 3 Make a hole through the rubber sole of each shoe, drilling from the heel to the toe. Widen the holes to fit a USB female socket at both heels, an accelerometer near the toe of the left shoe, and a photocell near the toe of the right shoe.

STEP 4 The left shoe will function as your bass drum pedal, while the right will work as a hi-hat pedal. Follow the circuitry diagram to attach their components.

STEP 5 Use the male-to-male USB connectors to attach both the drumsticks and the shoes to the remaining four USB female connectors. Then attach these USB ports to the Arduino UNO and attach the drumsticks and shoes to the five-pin DIN connector according to the diagram.

STEP 6 Download the drum kit code from popsci.com/thebigbookofhacks. Program your Arduino with the code and start running it.

STEP 7 Plug your MIDI device into the DIN port, then connect the MIDI device to your computer. Plug the 5-volt DC-power supply into a wall outlet.

STEP 8 Slip the shoes onto your feet, pick up your sticks, and drum away. Head banging is encouraged.

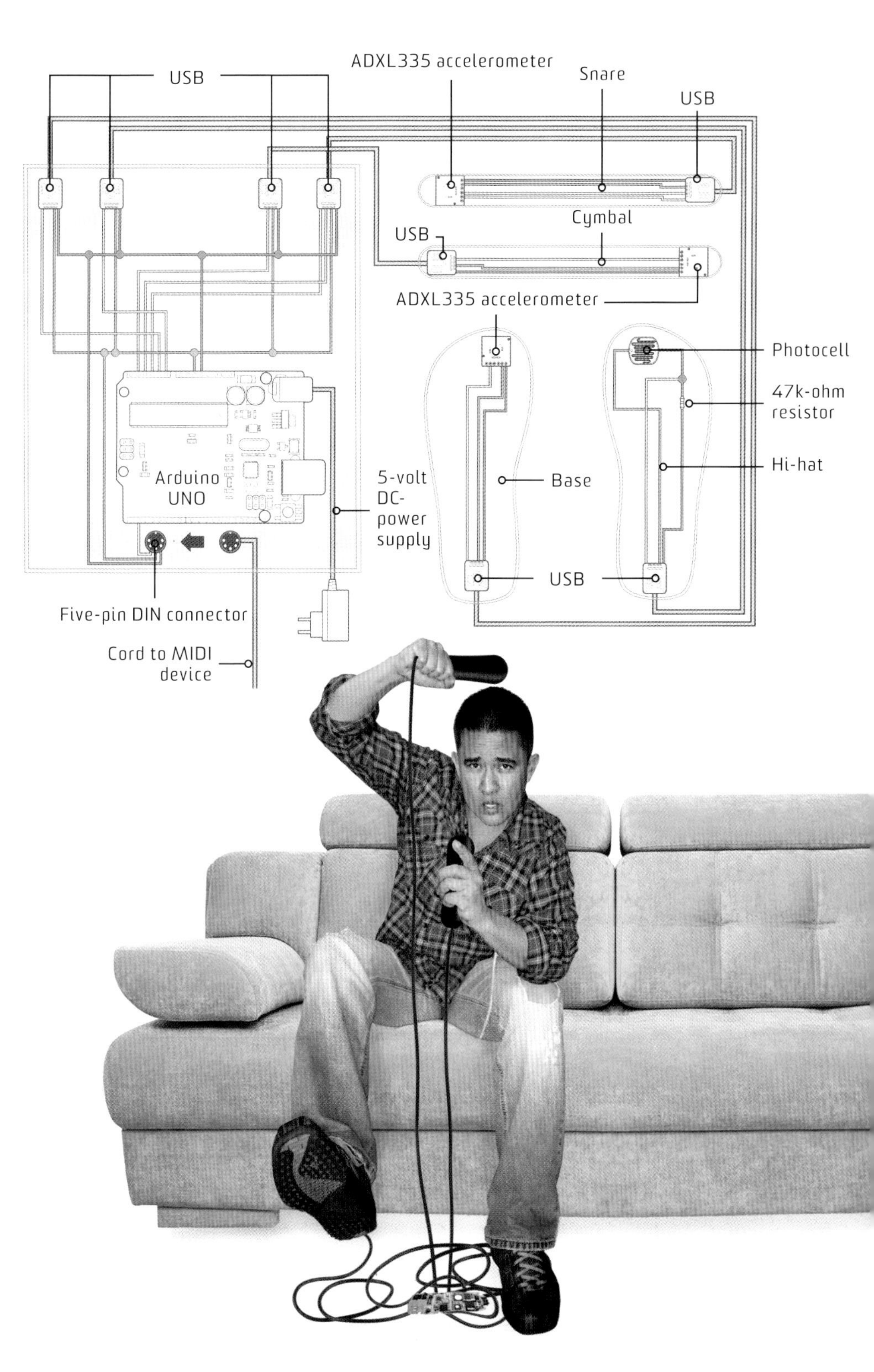

USB
ADXL335 accelerometer
Snare
USB
Cymbal
USB
ADXL335 accelerometer
Photocell
47k-ohm resistor
Hi-hat
Base
Arduino UNO
5-volt DC-power supply
USB
Five-pin DIN connector
Cord to MIDI device

5 MINUTE PROJECT

038 AMPLIFY MUSIC WITH PAPER CUPS

STEP 1 Turn one paper cup upside down and stick a toothpick through the bottom.

STEP 2 Put another cup on top of it so that it rests perpendicular to the first cup and is secured by the toothpick.

STEP 3 Cut a hole into the bottom of the top cup, and insert an earbud from your media player.

STEP 4 Repeat with the other two cups, then listen to your favorite tunes in stereo.

039 FILE-SHARE WITH A USB DEAD DROP

Camouflage a USB flash drive so that you can swap files on the sly.

MATERIALS

- USB flash drive
- Screwdriver
- Plumber's tape
- Drill with masonry bit
- Cement
- Paint, if desired

STEP 1 Stick your USB drive into your computer's port and upload any files you want to share, then remove it.

STEP 2 Scout for a good place to put your dead drop. You may need a drill with a masonry bit to make a hole in concrete, like the one you see here.

STEP 3 Slide your USB inside and use cement to secure it in place. Don't get any cement on the USB itself. If you want extra camouflage, paint a bit around it.

STEP 4 Scram—and let your contacts know where the secret docs are. To retrieve files from a dead drop, just line your laptop's USB port up with the USB drive and slide them together.

040 PIRATE A VINYL RECORD

If you live in fear of scratching a super-rare record, this silicone mold is for you.

MATERIALS

Four wood boards 14¼ inches (36.5 cm) in length
Nails
Hammer
Glass plate
Caulking
Record
Dowel
Silicone rubber designed for mold making
Casting resin

STEP 1 Nail together the boards to make a square wood frame. Place the frame on the glass plate, and seal around the inside edge with caulking.

STEP 2 Put the record you want to copy inside the frame on the glass plate—the side you want to copy should be face up. Fit a dowel into the record's hole.

STEP 3 Prepare the silicone rubber and pour it into the mold. Let it dry overnight.

STEP 4 Peel off the silicone mold, then trim any excess from around its edges.

STEP 5 Mix the casting resin and pour it into the silicone mold. Once it's set, loosen the cast and remove.

STEP 6 Pop your repro record onto your record player and hit play.

041 PLAY A POCKET THEREMIN

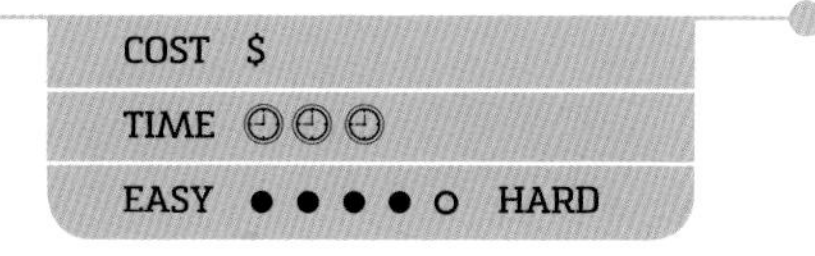

Throw together this pint-size, light-sensitive theremin for spooky sound effects on the cheap.

Remember those eerie sci-fi soundtracks from the 1950s? Chances are those oscillating noises were generated by a theremin. Designed by Russian physicist Léon Theremin and popularized by Robert Moog, a full-fledged theremin will set you back $400. Or you can build this pocket-size version—it's light-sensitive, so play it in subdued lighting for the best sound effects.

ROCK ON
Léon Theremin, inventor, plays his creepy-sounding instrument circa 1919.

MATERIALS

- Two 555 IC timers
- Two photocells
- Two 0.01-mF capacitors
- 1k-ohm resistor
- 5k-ohm potentiometer
- 2-position PCB terminal
- 8k-ohm, 1-inch (2.5-cm) speaker
- 9-volt battery snap
- Electrical wire
- Soldering iron and solder
- Drill
- Project box
- 9-volt battery

STEP 1 Wire your circuit according to the circuitry diagram at right, soldering your connections. Keep the wiring loose enough that you can insert it into the box.

STEP 2 Drill nine holes in the sides of the project box, spaced so that they match the circuitry diagram.

STEP 3 Insert the circuit inside the project box. Thread the loose wires through the holes in the box.

STEP 4 Solder on the photocells, potentiometer, speaker, and battery snap.

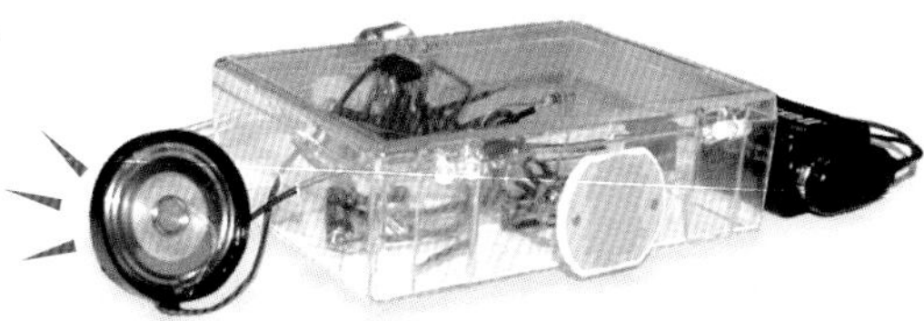

STEP 5 Connect the snap to the 9-volt battery. You should immediately hear an eerie noise. If you don't, check your wiring for faulty connections.

STEP 6 Head to a spot with low light. To produce a wide variety of sounds, move your hands over the theremin's photocells, which will vary the frequency and pitch of the output.

STEP 7 Now go film your own retro sci-fi flick; you've already got the sound effects in your pocket.

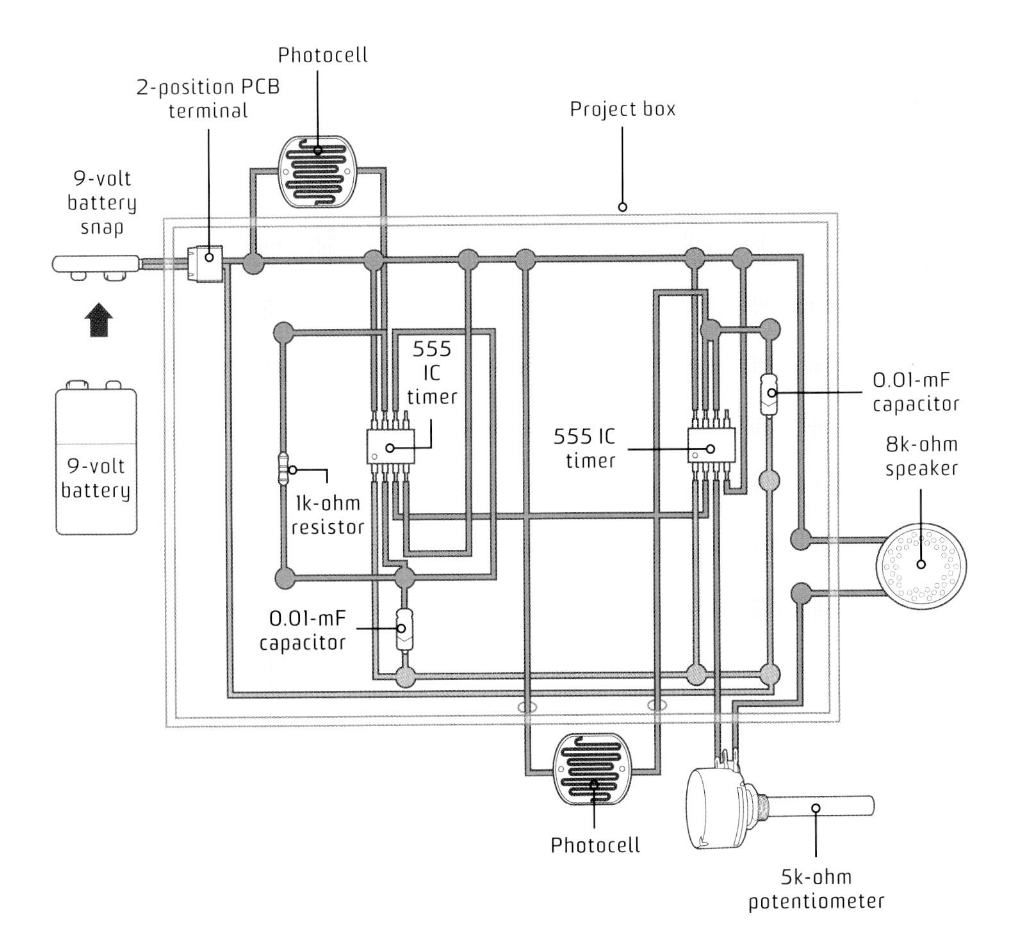

QUICK HACKS

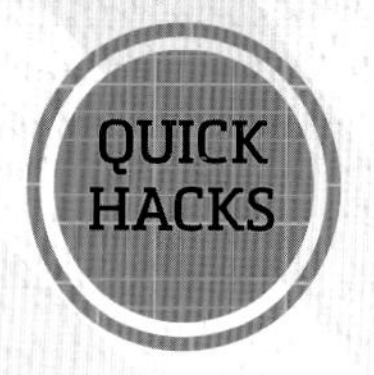

042 PUT A NEW SPIN ON AN OLD CD

In the age of MP3s, most people have a lot of old CDs lying around. Here's what to do with them.

EASY SPINNING TOP
Using a hot-glue gun, secure a large marble to the underside of a compact disc, right under the hole. Glue a plastic bottle cap to the top and give it a whirl.

BEER SPILL BLOCKER
Put a CD over your beer bottle so the bottle's neck sticks up through the CD's hole. Now when you knock the bottle over accidentally, the CD will prevent it from tipping all the way over—and spilling your brew.

ULTIMATE (COMPACT) DISC GOLF
This one's truly easy: Take an old CD and throw it around a disc golf course with some friends. Just don't throw it at your friends.

SUPERSHINY COASTERS
Cover compact discs with felt and use them as coasters. Make sure you cover the side with the artist's information on it—you don't want anyone knowing you once paid actual money for that Third Eye Blind CD, do you?

AIR HOCKEY IN A PINCH
Place a CD on a table about the size of, well, an air hockey table, and mark goal zones with tape. Stand across from your opponent, seize a CD spindle, and use it as an air hockey mallet to swat the CD back and forth.

043 CREATE AUDIO ART OUT OF CASSETTE TAPE

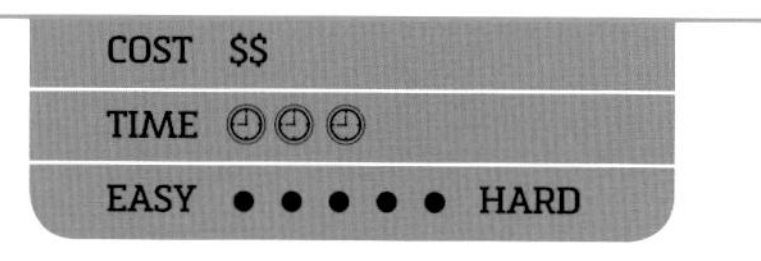

Get your John Cage on with a hack that's half musical instrument, half graffiti.

MATERIALS

- Cassette player and recorder
- Screwdriver
- Wire strippers
- Electrical wire
- Scissors
- Rubber thimble
- Superglue
- Glove
- Project box
- Drill
- On/off switch
- Soldering iron and solder
- Battery and battery holder
- Velcro
- Strips of audio tape

STEP 1 Pop open the cassette player's door. Push the play button to make a metal mechanical component called the *playhead* slide down. Remove the screws securing the playhead.

STEP 2 Turn your attention to the cassette player's back. Remove all the screws and open up the case. Take a good long look at the circuit board, noting where the playhead, speaker, and battery leads connect.

STEP 3 Unscrew and extract the circuit board with the speaker and playhead attached. Detach the drive motor and microphone from the circuitry itself.

STEP 4 Find the contacts that move together when you hit the play button. Sever the wires connecting them to the play button, but not the ones to the circuit board.

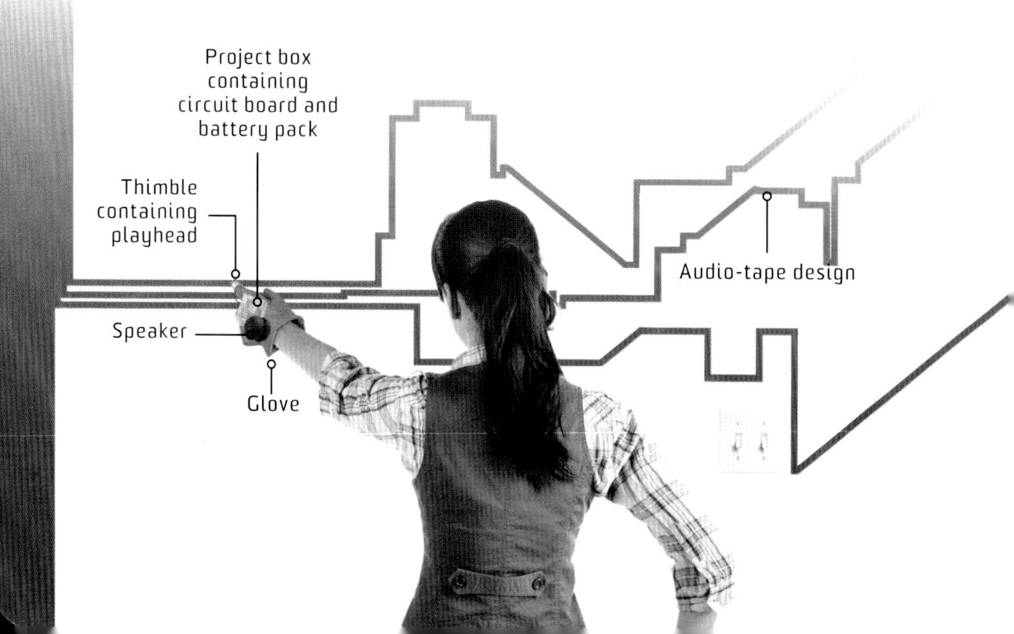

STEP 5 Cut the wires that connect the playhead and speaker to the circuit board.

STEP 6 Cut holes in the rubber thimble and thread the playhead's wires through them. Glue the playhead's base to the end of the thimble, and then glue the thimble over the finger of a glove.

STEP 7 Fit the circuit board into the project box, and drill holes into one side to thread the playhead's wires through. Reattach the wires to the circuit board.

STEP 8 Drill two holes in the project box for the on/off switch's wires. Solder these to the wires that the play button formerly controlled (they were connected to the contacts removed in step 4). Glue the switch to the box.

STEP 9 Cut a larger hole on top of the project box and super-glue the tape recorder's original speaker in the box. Reconnect the speaker's wiring to the circuit board.

STEP 10 Put the battery holder inside the project box. Attach the holder's wires to the circuit board where its original battery terminals were connected.

STEP 11 Attach the project box to the back of the glove with Velcro. The wire leading to the playhead on your fingertip should allow for hand and finger movement. If necessary, splice in extra electrical wire.

STEP 12 Take apart a cassette tape and remove its tape from the spools. Arrange the tape on an interior wall.

STEP 13 Place batteries into the battery holder, turn the switch on, and run the playhead over strips of audio tape. Experiment with speed—nail the right tempo, and you'll be able to hear the original recording.

5 MINUTE PROJECT

044 SCRATCH A PIZZA-BOX TURNTABLE

+ + + + + =

STEP 1 Cut a small hole in your pizza box's lid.

STEP 2 Tape your optical mouse inside the lid so that its eye points up through the hole when you close the box.

STEP 3 Cut a cardboard disc and attach it to the box using a pushpin so that it can spin around over the mouse's eye.

STEP 4 Open your mixing software and scratch away.

045 USE A BIKE-PART SPIROGRAPH

Those mathy whirls of color from your childhood can be yours all over again.

MATERIALS

- Bike chain
- Thin plywood
- Tape
- Pen
- Jigsaw
- Superglue
- Colored pens
- An assortment of chain rings

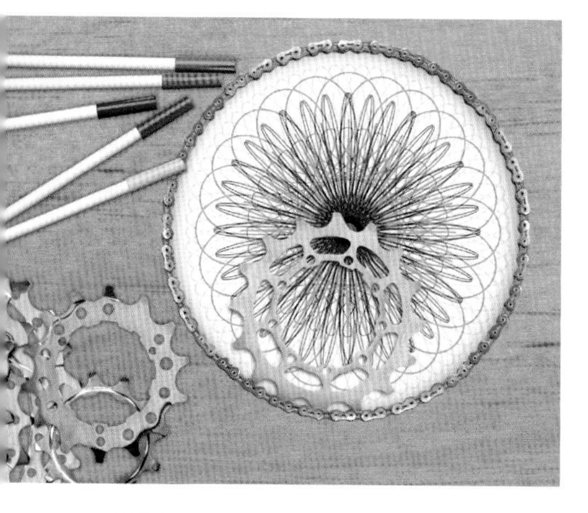

STEP 1 Arrange a bike chain on the plywood in a perfect circle and tape it down. Faithfully trace around the chain's outer rim, being careful not to move it.

STEP 2 Using a jigsaw, cut out the circle and discard it. Line up the chain so that it fits inside the circular hole.

STEP 3 Use superglue to secure the chain all the way around the inner edge of the hole in the plywood. Let it dry overnight.

STEP 4 Place the plywood frame over paper. Put a chain ring on the paper and insert a colored pen through one of its bolt holes to make contact with the paper.

STEP 5 Keeping your pen in the chain ring's bolt hole, trace around the chain to lay down a design. Experiment—chain rings with different numbers of teeth will create patterns of varying complexity.

5 MINUTE PROJECT

046 TURN JUNK MAIL INTO PENCILS

STEP 1 Cut a strip of paper—with colors or a pattern that you like—so it's 16½ inches (42 cm) in length and 5 inches (12.5 cm) in width.

STEP 2 Apply glue along a long edge and apply a piece of mechanical pencil lead to it, then roll your paper snugly around the pencil lead.

STEP 3 Continue applying glue every 1 inch (2.5 cm) and rolling up the paper little by little.

STEP 4 Let the pencil dry overnight. Use a craft knife to sharpen it and start scribbling.

047

SET UP A TURNTABLE ZOETROPE

See LEGOs come to life with this classic animation trick.

COST $$

TIME ◷◷

EASY ● ● ○ ○ ○ HARD

The world saw its first modern zoetrope in 1833, and since its invention the device has paved the way to cinema as we know it. This playful update uses a strobe light to interrupt your view of a series of still objects as they go around and around on a record player—causing your eye to perceive them as if they were in motion. It's not 1833 anymore, but the effect is still pretty mind-boggling.

MATERIALS

- Protractor
- A record to sacrifice
- 18 LEGO miniature figures
- Superglue
- Record player
- Strobe light

STEP 1 Using the protractor, measure and draw lines every 20 degrees on the sacrificial record. Space out the LEGO figures around the edge of the record according to these marks, and glue them down.

STEP 2 Put the LEGO figures in positions of your choosing—think about creating the look of continuous motion by carefully changing each one's position incrementally from that of the one before it.

STEP 3 Set the record player to 33$^{1}/_{3}$ RPM.

STEP 4 Adjust your strobe light to flash ten times per second and position it so that it's pointed at the zoetrope at close range.

STEP 5 Turn out the lights, turn on the record player and the strobe, and watch those LEGO figurines start dancing, running, battling, or doing whatever else you want to make them do.

048 WIELD A POTATO LAUNCHER

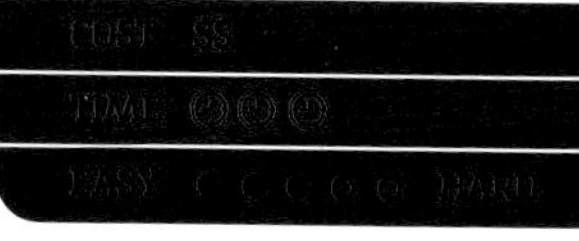

Potatoes—they're not just for the dinner table. Build this cannon and see spuds fly.

MATERIALS

- 2 feet (60 cm) of 16-gauge insulated wire
- Wire strippers
- BBQ igniter
- Various PVC parts (see diagram below)
- Soldering iron and solder
- PVC primer
- PVC pipe cement
- Drill
- Two 8-by-2½-inch (20-by-6.35-cm) machine screws
- Leather gloves
- Electrical tape
- Hair spray
- Potatoes

STEP 1 Cut the insulated wire in half and strip back a bit of the ends of each piece.

STEP 2 Find the fine wire near the igniter button's lip. Twist this wire's end with one stripped insulated wire; solder them together. Secure with electrical tape.

STEP 3 Locate the igniter's main wire near the base. Cut it, leaving about 2 inches (5 cm) at the plug end. Strip, twist, and solder the end with the other insulated wire.

STEP 4 Prime the adapter, coupler, combustion chamber, and bushing. (Don't get primer on the threads.)

STEP 5 Immediately apply pipe cement to the parts of the adapter, coupler, combustion chamber, and bushing that will fit together. Cement the coupler and bushing to the combustion chamber, and the adapter to the coupler.

STEP 6 Right the assembled pieces and twist them while pressing, then check that the combustion chamber fits 1½ inches (3.75 cm) into the coupler and equally far into the bushing.

STEP 7 Wait 10 minutes for the cement to dry, then drill two holes in the

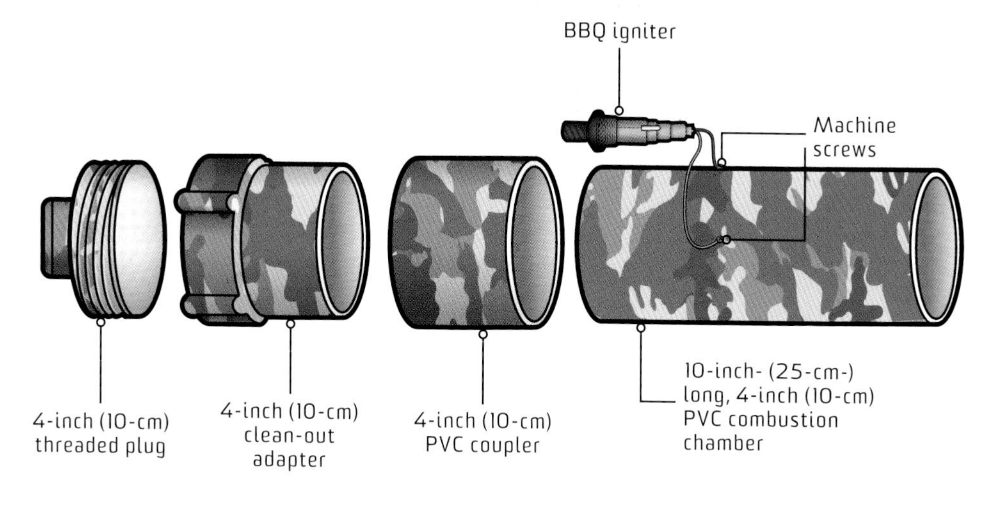

WARNING
You're building a pretty powerful spud gun here, buddy. Be careful of flammable fuels and always, always watch where you point this thing.

combustion chamber at a 90-degree angle to one another, closer to the coupler than the bushing. Drive in two machine screws, leaving 1/4 inch (6.35 mm) between them inside the pipe.

STEP 8 Prime one end of the 36-inch (90-cm) pipe—this is your barrel—and the smaller, exposed end of the bushing. Apply pipe cement and twist to seal.

STEP 9 Wrap the ends of each of the insulated wires around the screws in the combustion chamber. Then tighten the screws and insulate with electrical tape.

STEP 10 Wearing leather gloves, lash the ignitor button to the combustion chamber's side using electrical tape.

STEP 11 Let the contraption dry for 48 hours before using it. (Otherwise, it may blow up—trust us on this stuff.) Test the igniter button to make sure there's a spark. If there is, twist on the threaded plug at the end.

STEP 12 Place the launcher on the ground and securely lodge a potato into the end of the barrel, about 2 inches (5 cm) inside. Rotate the potato to mold it into a cylindrical shape that fits tightly inside the barrel.

STEP 13 Remove the end cap and spray a 2-second blast of hair spray into the chamber. Close it again.

STEP 14 Pick up the launcher, point it in a safe direction, and press the BBQ igniter. Watch that tater go!

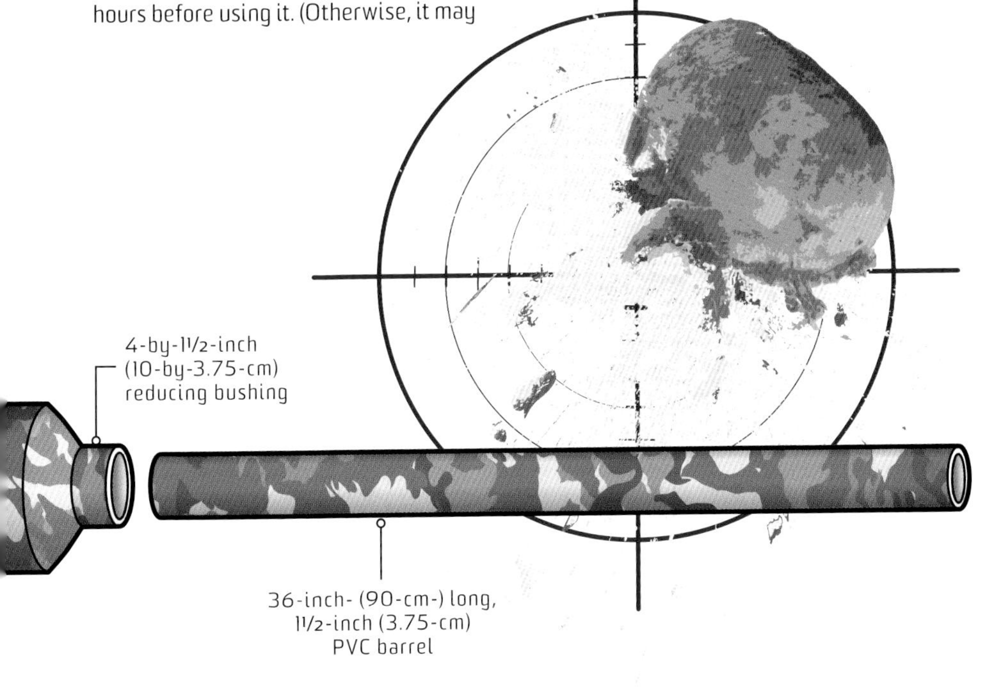

049 THE REAL IRON MAN SUIT

This homebuilt superhero suit looks as good as the silver-screen version.

Anthony Le has been a fan of Iron Man since he was a kid. When he heard that the comic-book superhero was hitting the big screen in 2008, he was inspired to build his own Iron Man suit. That version was more of a costume, but his next edition—finished just in time for the movie's sequel—edges much closer to the real thing. With its dent-proof exterior, motorized faceplate, and spinning mock Gatling gun, his take on the movie's War Machine suit could easily frighten a supervillain—not to mention the kids in the theater.

To make the suit, Le studied some concept sketches of the suit posted on the Internet. He used thin, high-impact urethane for the armor, cutting it into plates and joining them with some 1,500 rivets and washers. He sculpted a clay helmet mold and used a mix of liquid resin to create the final product. He added a replica of the machine gun on the suit's shoulder made of PVC pipes and other materials. He also added a servo motor that opens the faceplate, and built a gun out of pipes and a motor. LEDs in the eyes and chest-plate further add to the illusion.

The LEDs and motors that drive the gun and the faceplate have batteries hidden in the suit's large frame. Le added a hands-free button that activates the helmet. When the faceplate is open, he just stands up and points his arm forward, causing his chest to press against the button, triggering the servo motors in the helmet to close the mask. This, in turn, switches on the red LEDs set inside the eye openings, which are large enough for him to also see out of. To open the mask up again, he presses another button.

HARD TO MISS

The material Le used for the armor is thin but takes stress well. "You can throw it against the wall, and it won't even be damaged," he says. Le has a cult following among fans of the movie featuring Iron Man—he wore the suit to the theater to see *Iron Man 2*.

5 MINUTE PROJECT

050 MIX MAGNETIC SILLY PUTTY

BORAX

STEP 1 Mix 1 tablespoon of basic craft glue and 1 cup (240 ml) of water in a plastic bag, then add 1 tablespoon of borax. Squeeze to make the putty.

STEP 2 Wearing gloves and a face mask, spread out the putty and sprinkle about 2 tablespoons of iron oxide powder onto it. Work it in for about 5 minutes.

STEP 3 Introduce a magnet to your putty and watch it move and change shape.

051 COOK UP FERROFLUID

Believe it or not, this spiky stuff is a fluid. Put it on a magnet and it goes nuts.

MATERIALS

- Syringe
- Ferric chloride solution
- Distilled water
- Steel wool
- Coffee filter
- Household ammonia
- Oleic acid
- Kerosene

STEP 1 With a syringe, measure 10 ml of ferric chloride solution and 10 ml of distilled water into a container.

STEP 2 Add a small piece of steel wool. Stir or swirl the solution until it turns bright green, then filter it through a coffee filter.

STEP 3 Add 20 ml more ferric chloride solution to your filtered green solution. While stirring, add 150 ml of ammonia.

STEP 4 In a well-ventilated area, heat the solution to near boiling. While stirring, add 5 ml of oleic acid. Keep heating until the ammonia smell is gone (about an hour).

STEP 5 Let cool, and then add 100 ml of kerosene. Stir until the black color attaches to the kerosene.

STEP 6 Pour off and collect the kerosene layer in a bowl.

STEP 7 Place a magnet under your bowl's bottom and watch the weird peaks rise.

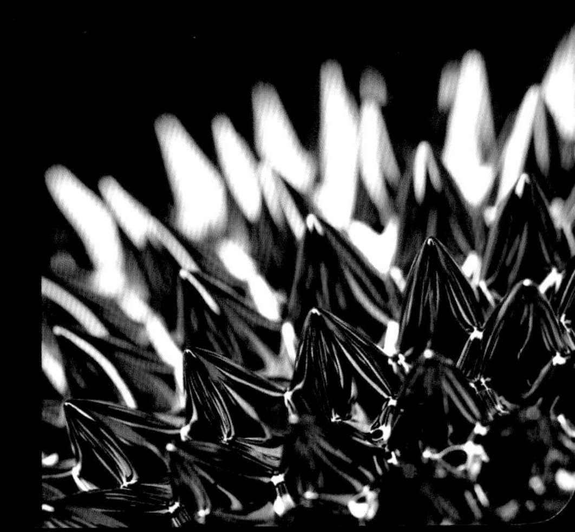

052 CATCH A THRILL ON A BACKYARD COASTER

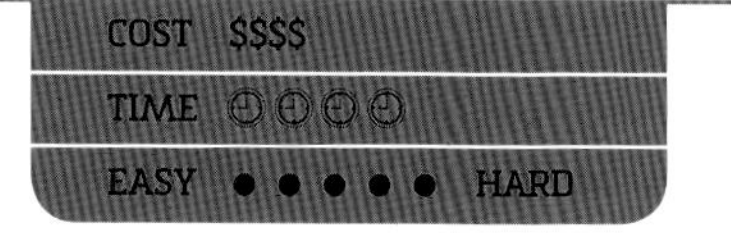

Sure, it may not do loop-de-loops, but our lawyers tell us that's a good thing.

MATERIALS

- 5-by-5-inch (12.5-by-12.5-cm) wooden posts
- Drill
- Screws and bolts
- 4-by-4-inch (10-by-10-cm) wooden posts
- 2-by-4-inch (5-by-10-cm) boards
- Measuring tape
- Circular saw
- Gray sun-resistant 1½-inch (3.75-cm) PVC pipe
- Heat gun
- PVC of a slightly smaller diameter that fits snugly inside the main PVC size
- Casters fitted with skateboard wheels
- Rectangular piece of wood or particleboard
- Car seat with seat belt
- Rope

STEP 1 Survey your land and choose a spot for your backyard coaster.

STEP 2 Design your track. Note that you can't have a hill that is higher than the previous peak. If you're creating multiple hills, the car must continuously build enough energy by descending hills to make it up subsequent hills and around corners. Try starting with a downward slope from about 9 feet (2.75 m) followed by a hill that measures about 6 feet (1.8 m) at its crest. End the track with a steep uphill that the car won't be able to climb over, which will stop it.

STEP 3 Lay down 5-by-5-inch (12.5-by-12.5-cm) posts on the ground as a base. Screw them together with a drill.

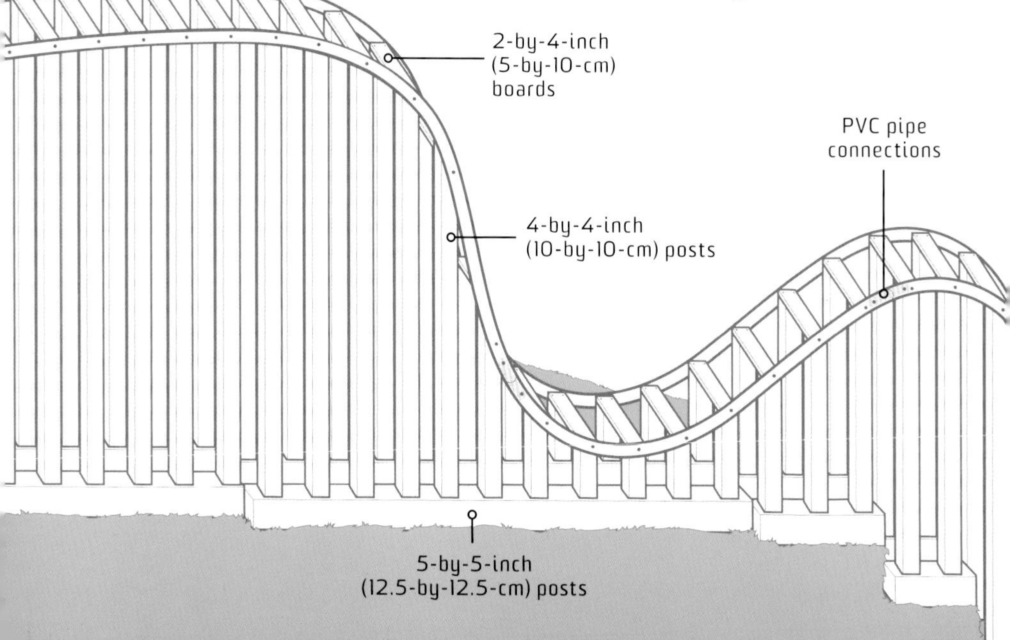

WARNING
You've seen the movie *Jackass,* right? You know that part where they tell you not to perform any of the stunts seen in the movie? That goes for this, too.

STEP 4 Saw the 4-by-4-inch (10-by-10-cm) posts to varying heights that match your design's elevation changes. Then screw the posts along the top of the 5-by-5-inch (12.5-by-12.5-cm) floor to create your elevation. Leave 1 foot (30 cm) between each post.

STEP 5 Mark and cut 2-by-4-inch (5-by-10-cm) boards into slats as long as the width of your track. Lay these across the tops of the 4-by-4-inch (10-by-10-cm) posts and screw them into place.

STEP 6 Drill holes through the gray PVC pipe, and then use the drill to screw through these holes into the sides of the 2-by-4-inch (5-by-10-cm) boards, forming the rails and sides of your track. Use a heat gun to mold the PVC to curves. As you go, connect segments of PVC securely by sliding a 1-foot (30-cm) section of the smaller diameter pipe into the PVC that's already in place. Screw it in, then slide on the new piece and screw that in as well.

STEP 7 To make the coaster car, screw wheels onto the bottom of a flat piece of wood that's 3 feet (90 cm) long and about the width of the track. One set of wheels should roll on top of the PVC and the other should ride along the outside.

STEP 8 Check to see if the car works by rolling it on the track, and make adjustments to wheel placement if necessary.

STEP 9 Bolt the car seat equipped with a seat belt onto the coaster car, leaving room in front for the rider's tucked legs.

STEP 10 Attach a long piece of rope to the car—you can use this to pull the coaster backward up the first slope so that a rider can board.

STEP 11 Test the roller coaster. Place the car on the track and put a sack of flour on it to serve as a crash-test dummy. Test more than once. If the tests are successful, consider someday cautiously trying it yourself.

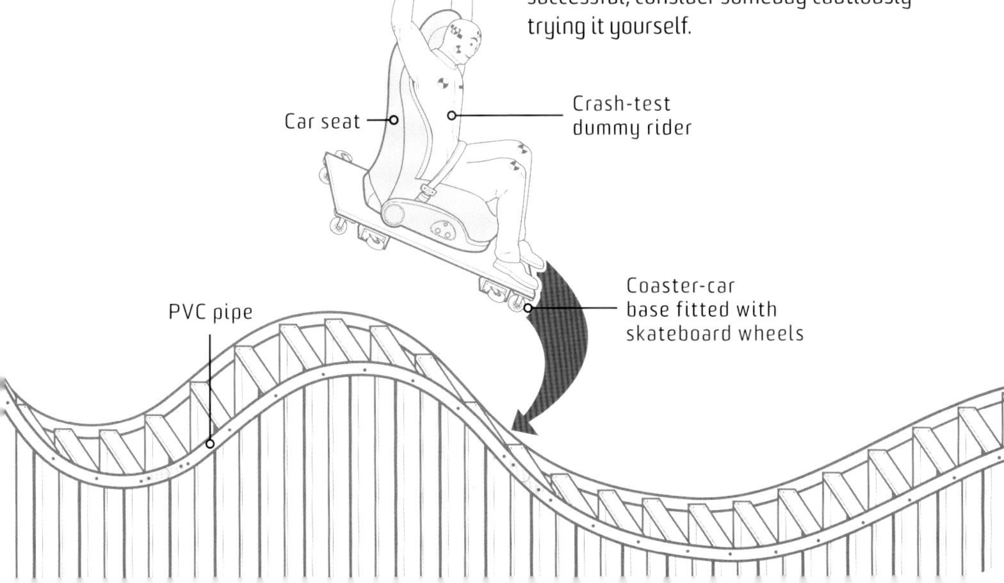

053 SET UP A PINBALL GAME AT HOME

COST	$$
TIME	◷◷◷
EASY	● ● ● ○ ○ HARD

Because arcades can't stay open all the time.

MATERIALS

- Peg-Board
- Saw
- 2x4s
- Wood glue
- Nails
- Hammer
- Drill
- 3/4-inch (2-cm) drill bit
- 1/2-inch (1.25-cm) drill bit
- 1/8-inch (3-mm) drill bit
- 5/16-inch- (8-cm-) diameter hex bolt with nut
- 3/8-inch- (9.5-mm-) diameter spring
- Wood
- Wood pegs
- Bicycle bell
- Rubber bands
- Scrap wood
- PVC pipe elbows
- Rubber hose
- Marbles

STEP 1 Use the saw to cut the Peg-Board to your desired size to form the base of the pinball machine.

STEP 2 Measure and cut 2x4s to make a frame. Glue and nail it into place.

STEP 3 Nail a 2x4 under the top of the frame to prop it up at a slight angle.

STEP 4 Drill a hole about 1/2 inch (1.25 cm) in diameter in the bottom frame piece's right corner. Insert the hex bolt and slide the spring over the bolt.

STEP 5 Attach the nut to secure the spring on the bolt. Pull the bolt down to compress the spring—this is the ball launcher. Place a piece of scrap wood alongside the bolt to create a guide for the ball.

STEP 6 Toward the bottom of the game, drill through the side boards to create a 3/4-inch (1.9-cm) hole in each. Drill another hole next to the first so that the holes meet, making a long oval hole on each side of the board.

STEP 7 Mark the center of the oval hole on the top of the board, then drill into it from above with a 1/8-inch (3-mm) drill bit. Repeat on the other side.

STEP 8 Cut two pieces of wood into paddles of your desired length. Sand the edges of the paddles.

STEP 9 Drill a 1/8-inch (3-mm) hole from the top near the ends of the paddles. Then slip the paddles into the holes you made in the side boards.

STEP 10 Place a nail through the holes in the frame, through the paddles, and down into the holes' bottoms. Tap the nails with a hammer to secure them in place. Then place a peg on either side of each paddle to restrict its range of movement.

STEP 11 To create tunnels, nail down rubber hoses sliced in half lengthwise and PVC pipe elbows, and for good bumper action, extend rubber bands between pegs or nail scrap wood to the Peg-Board. If you want spinners, try foam X shapes secured loosely with a nail. Don't forget a ramp and bicycle bell.

STEP 12 Load up the launcher with a marble, pull back the bolt, and release it—let the game begin.

FROM THE ARCHIVES

054 Play DIY Skee-Ball

Go analog with a good old-fashioned Skee-Ball toss.

MATERIALS

- 1-inch (2.5-cm) particleboard
- 1/2-inch (1.25-cm) particleboard
- 1/4-inch (6.35-mm) particleboard
- Saw
- Wood glue
- Nails
- Rubber ball

STEP 1 Use the saw to cut pieces of particleboard according to the measurements in the diagram.

STEP 2 On the board that serves as a ramp, trace your ball at one end to create a circular shape. Then add 1/4 inch (6.35 mm) all around this shape, and cut it out.

STEP 3 Build the game using nails and wood glue, making sure that the ramp fits against the back wall and that the opening is large enough for the ball to drop in.

STEP 4 Cut a hole into one side of the box so that you can reach in and grab the ball after scoring tosses.

STEP 5 Stand about 8 feet (2.4 m) away and toss the ball so that it rolls up the ramp—if it goes in the hole, that's one point. Any player who scores may continue until he or she misses; the first to score 15 points wins.

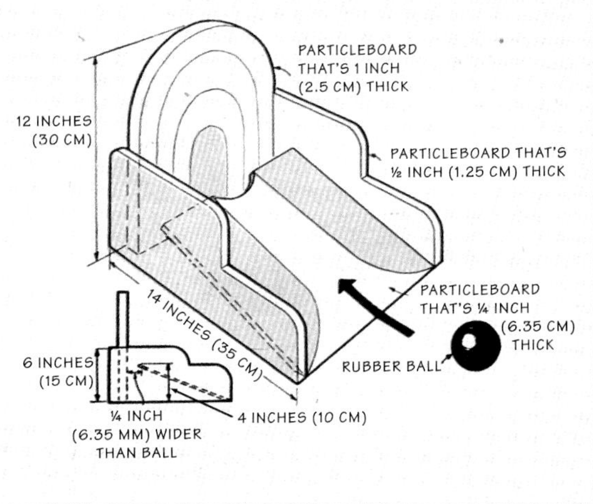

055 BUILD A MINI ARCADE GAME

No room for a full-on console? Make a mini version.

MATERIALS

- Netbook
- 1/2-inch (1.25-cm) medium density fiberboard
- Ruler
- Screws
- Screwdriver
- Hinge
- Hole saw
- Buttons and joystick
- Label paper
- Paint
- Table saw with a plastic-cutting blade
- Plexiglas
- I-PAC 2
- Electrical wire
- Soldering iron and solder
- Hot-glue gun
- USB cable
- Legal arcade software

STEP 1 Take apart your netbook and measure its LCD screen, obtaining the dimensions on which you'll base your arcade design.

STEP 2 Measure and cut ten pieces of medium-density fiberboard for the arcade's body (one each for the back, top, bottom, and sides) and five pieces for the front.

STEP 3 Screw it together. Mount a hinge in the back so you'll be able to open it up and access your electronics.

STEP 4 Measure and cut holes in the front of the arcade for the buttons and joystick using the hole saw.

STEP 5 Come up with a design for your arcade and print it on the label paper. Paint the arcade your desired color, then apply your label to the arcade.

STEP 6 Use the table saw with a plastic-cutting blade to cut Plexiglas pieces that fit over the pieces of board, and screw them over the label paper to protect the design.

STEP 7 Place the buttons and joystick, and then wire them up according to the I-PAC 2's instructions.

STEP 8 Cut an additional piece of fiberboard to mount behind the LCD, propping the LCD up inside of the arcade. Make a hole in this board for wires.

STEP 9 Use the table saw to cut out a piece of Plexiglas. Then mount the piece in front of the LCD.

STEP 10 Mount the I-PAC 2 and the base of the netbook inside the back panel of the arcade with hot glue.

STEP 11 Solder the LCD wire and USB wire directly to the I-PAC 2 board.

STEP 12 Open the arcade's back and turn on the netbook. Download and install some legal arcade software, and get your game on.

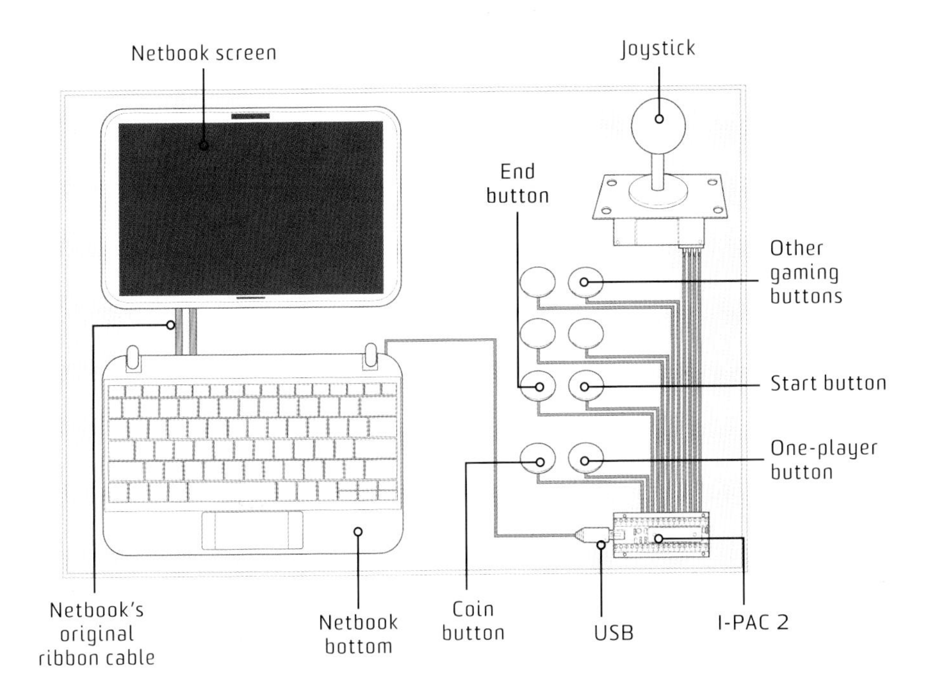

5 MINUTE PROJECT

056 CINCH A NINTENDO-CONTROLLER BELT

STEP 1 Use a screwdriver to open up the back of the controller and cut off the cord where it connects to the circuit board. Reassemble the controller.

STEP 2 Cut the adhesive Velcro to fit the surface of the belt buckle.

STEP 3 Adhere one piece of Velcro to the buckle and the other piece to the back of the controller.

STEP 4 Hold your pants up, nerd.

057 SET UP A SUPERSIZE GAME OF OPERATION

This classic kids' game may be all grown up, but it'll still leave you in stitches.

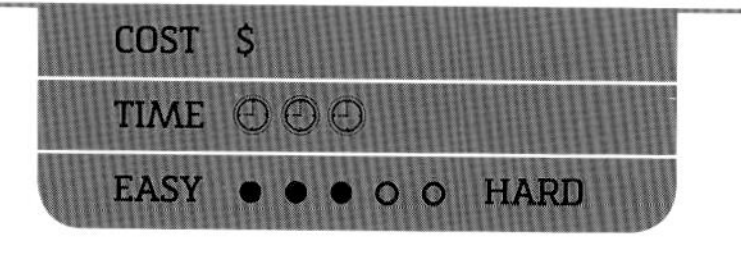

MATERIALS

- Large, flat box
- Paint
- Seven "body parts"
- Box cutter
- Disposable tin oven trays
- Tape
- Gray paper
- Wire strippers
- Buzzer
- Three AA batteries and holder
- Threaded electrical wire
- Five red LEDs
- Five 100-ohm resistors
- Soldering iron and solder
- Metal tongs
- Multimeter
- Clear ping-pong ball, cut in half

STEP 1 Paint your box and sketch an outline of the patient's body, adding as much detail as you want.

STEP 2 Outline the body parts you'll be "surgically removing." Add at least 1/2 inch (1.25 cm) around the outlines, then use a box cutter to cut them out. Cut a hole for the nose that can fit all five LEDs, and a hole in the side of the box so you can easily access its interior.

STEP 3 Cut up and/or combine the oven trays so that they fit closely around each body part, then tape them under the holes. Line the trays' bottoms with gray paper.

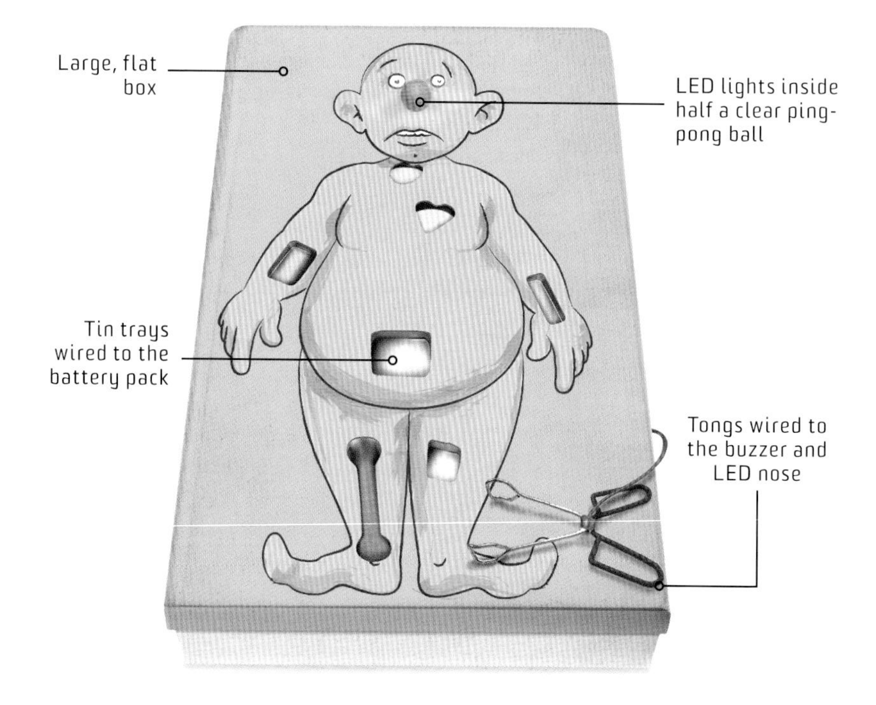

STEP 4 Strip plastic off the ends of the buzzer's wires and those of your battery holder, then connect the buzzer's positive end to the battery pack's positive end.

STEP 5 Twist each positive leg of the five LEDs to a resistor, then wire all five resistors to a length of threaded electrical wire. Solder this wire to the battery holder's positive wire.

STEP 6 Twist the buzzer's negative wire to each LED's negative leg. Then solder in place.

STEP 7 Cut a piece of electrical wire long enough to extend from underneath the "patient's" head to outside the box, where the players will pull it all over the board. Attach one end of this wire to the buzzer's negative wire.

STEP 8 Strip the protective coating on the tongs, then twist the other end of the wire that's attached to the buzzer's negative wire around the bare space. Tape the wire in place.

STEP 9 To wire the oven tins, cut a long electrical wire for each, making sure that it will reach the battery pack's location near the head. Tape each wire to an oven tin, then connect it to the battery pack's negative wire.

STEP 10 Tape the red LEDs in the hole for the nose and glue half of a ping-pong ball over the LEDs. Tape the battery pack and buzzer nearby.

STEP 11 Decorate your patient, drop body parts into the tins, and page your friends to come "operate."

058 Play Giant Checkers

Fact: If something is fun, making it larger makes it even more fun.

MATERIALS

Particleboard
Saw
Paint
U-shaped nails
Two pole hooks

STEP 1 Cut nine 4-foot- (1.2-m-) square pieces of particleboard, and arrange them in a 12-foot (3.6-m) square. (Using segments will allow you to store and transport the board.)

STEP 2 Mark a grid of squares that each measure 1 foot (30 cm). Paint the squares in alternating colors of your choice.

STEP 3 Use a saw to cut 24 discs that measure 8 inches (20 cm) in circumference. Paint half the discs one color, half another color.

STEP 4 Hammer a U-shaped nail partway into the center of each disc.

STEP 5 Set up the discs on the board. Use the pole hooks to pick up and move them by the nail loops.

059 BUILD A MINT-TIN RACER

STEP 1 Use a drill to make five holes into your tin: two on both of the tin's long sides and one in the top corner of the lid.

STEP 2 Measure and cut two wooden sticks so they are long enough to traverse the width of the mint tin with about 1/2 inch (1.25 cm) extra on either side.

STEP 3 Slide the axles through the holes in the side of the tin, and attach bottle-cap "wheels" to the sticks with hot glue.

STEP 4 To deck out your racer with a flag, insert a straw into the hole in the tin's top, and tape a triangular flag to the straw's top.

STEP 5 Detail your racer however you like.

060 SHAKE UP A MARTINI IN A MINT TIN

STEP 1 Drill a hole into one end of the mint tin and insert a plastic nozzle. (You can buy these in bulk online or at home-improvement stores.)

STEP 2 Fill two tiny bottles that fit inside the tin: one with gin or vodka, one with vermouth.

STEP 3 Place the booze bottles, a paper cup, and an olive inside the tin.

STEP 4 When you need an emergency drink, remove all the tin's contents and pour the bottles into the tin.

STEP 5 Close the tin and shake it well.

STEP 6 Loosen the nozzle, pour into the cup, and garnish with the olive.

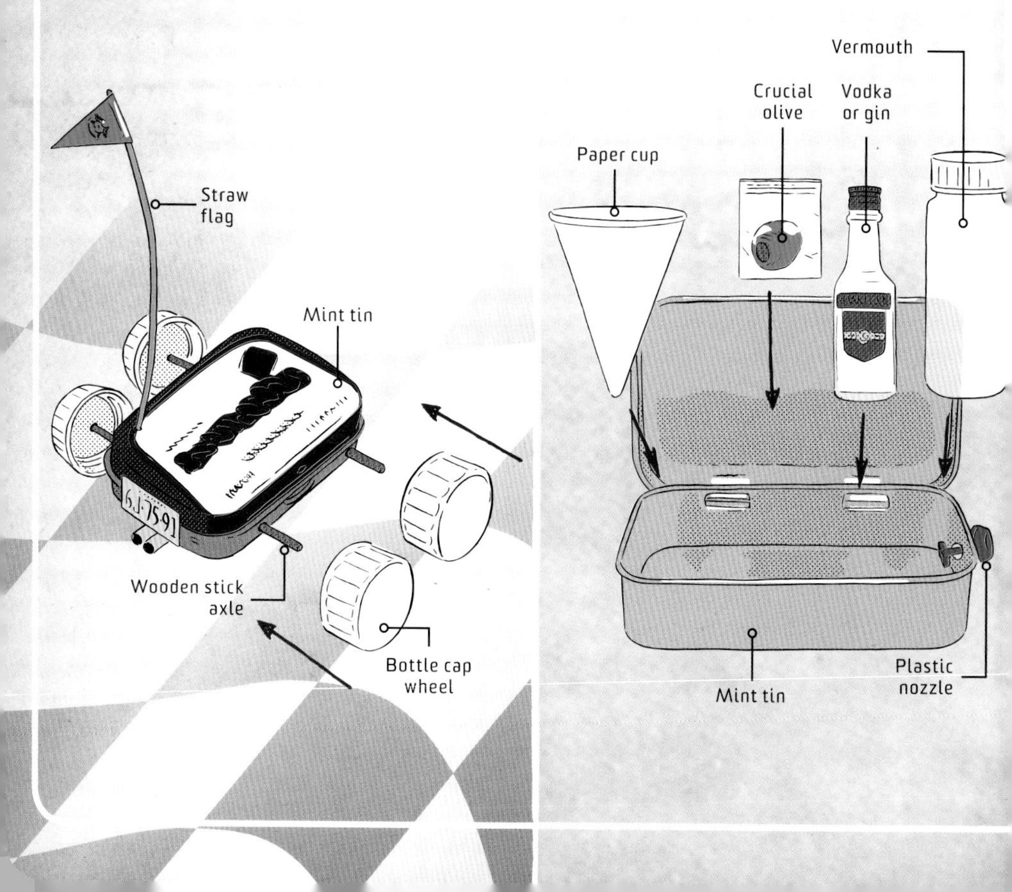

061 STRUM A MINT-TIN GUITAR

STEP 1 Position the tin so that its label is facing you. Then trace the stacked ends of three rulers onto the far right side of the tin to make a rectangle.

STEP 2 Using a drill, make a hole in the rectangle on the box's side, then cut out the rectangle outline with tin snips.

STEP 3 On the other side of the box, just below the lid's lip on the side, make three evenly spaced holes for the guitar strings. Thread the strings through and knot them off inside the box.

STEP 4 Remove the insides of a cheap ballpoint pen and cut the clear tube to about the width of your mint tin. Then cut it in half lengthwise. Make three notches in it for your guitar strings.

STEP 5 Use a hot-glue gun to glue the pen tube facedown onto the tin lid on the side where you've made holes for the strings.

STEP 6 Insert one ruler into the rectangular cutout so that it goes about halfway into the mint tin. Secure it with a hot-glue gun.

STEP 7 Cut a credit card to the ruler's width. Bend one edge up and glue it to the ruler about 1/2 inch (1.25 cm) from the ruler's end.

STEP 8 Cut the other two rulers down to 1 inch (2.5 cm) shorter than the exposed ruler. Glue them on top of the first ruler.

STEP 9 Drill holes for the eyebolts into the end of the bottom ruler. Insert and secure them with nuts on the ruler's bottom.

STEP 10 Run the strings over the pen tube, tie them off around the eyebolts, and start strumming the hits.

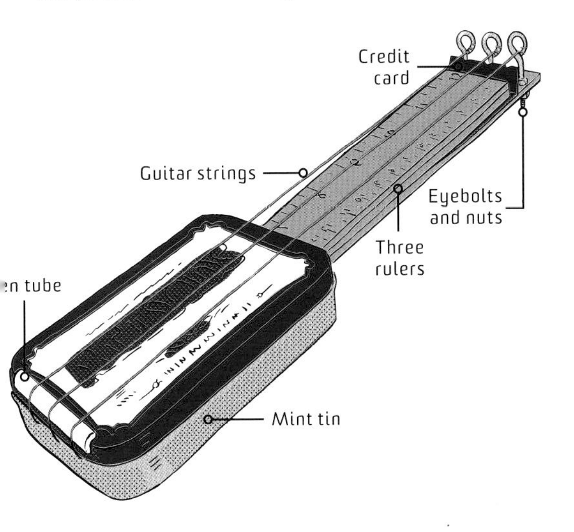

062 CARRY A POCKET BILLIARD SET

STEP 1 Use a craft knife to cut a piece of foam to fit inside a mint tin. It should be just about level with the tin's top.

STEP 2 Remove the foam and place it on green felt. Trace around it, then cut out the shape. Test it to make sure that it fits nicely inside the tin.

STEP 3 Take a small bead (aka, one of your pool balls) and place it in a corner of the felt. Cut a hole around it to make a pocket. Then trace the scrap to make pockets in the other three corners.

STEP 4 Glue the felt down onto the foam. Trace and cut the pocket shapes out of the foam, too.

STEP 5 Assemble the seven beads into a triangle. Place a piece of copper wire along one side of the triangle; mark its length.

STEP 6 Use this measurement to fold the wire into a triangle shape with sides of equal length.

STEP 7 Cut a small-diameter copper rod to make a pool cue.

STEP 8 Rack 'em up wherever you go.

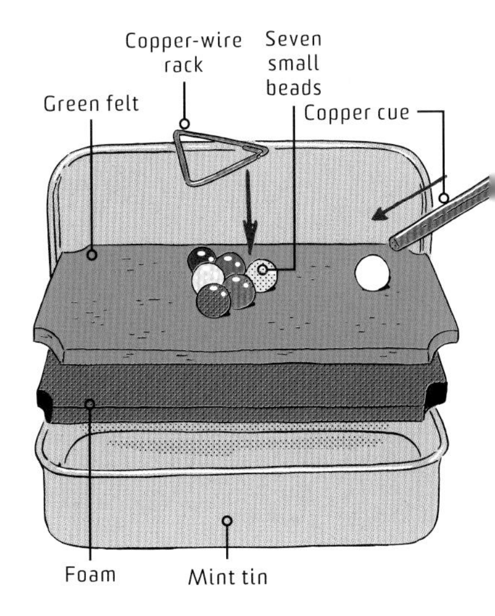

063 HACK YOUR MAGIC 8 BALL

Can you make this fortune-telling gizmo say whatever you want? It is decidedly so.

MATERIALS

- Magic 8 Ball
- Plexiglas cutter
- Screwdriver
- Razor
- Sandpaper
- Extra-fine permanent marker
- Superglue

STEP 1 Use a Plexiglas cutter to cut through the glue at the ball's equatorial seam, then carefully pry it open.

STEP 2 Inside there's a cylinder of blue dye. Remove the screws that hold it in place. Pour out and reserve the liquid.

STEP 3 Fish the answer ball out and pat it dry. Use a razor and sandpaper to scrape and sand the existing messages off the faces of the ball.

STEP 4 Write the messages you desire on the ball using an extra-fine permanent marker. Wait for the ink to dry.

STEP 5 Put the answer ball inside the cylinder, and then transfer the blue liquid back into the cylinder, too.

STEP 6 Glue the cap on the cylinder and reinsert it inside the two halves of the Magic 8 Ball.

STEP 7 Glue the ball back together, let it dry, and enjoy rewriting the future.

064 GO ANYWHERE WITH VIRTUAL-REALITY GLASSES

. . . anywhere that Google Street View goes, that is.

MATERIALS

Safety goggles
Large piece of cardboard
Pencil
Craft knife
Tape
Smartphone

STEP 1 Lay the safety goggles on the cardboard so that they're facing forward. Trace around the shape, adding at least 2 inches (5 cm) in front.

STEP 2 Roll the goggles up so that they're resting on an end. Trace around that side, adding extra space again.

STEP 3 Using the craft knife, cut out your tracing as one piece, then fold it so you have a four-sided rectangular tube that fits perfectly around your goggles. Secure it with tape and slide the goggles just inside.

STEP 4 On a separate piece of cardboard, trace the cardboard box's front, leaving 1-inch (2.5-cm) tabs on either side. Cut it out, then trace your smartphone onto its center. Cut out the shape of your phone, making a window, and insert this cardboard piece into the rectangular tube opposite the goggles.

STEP 5 Dial up Google Street View, locate a place you've always wanted to go, and tape your phone over the window.

STEP 6 Don your virtual-reality glasses, and take a walk someplace far, far away.

065 PUT ON A LIQUID LIGHT SHOW

Project extreme grooviness with this simple psychedelic setup.

MATERIALS

Cardboard sheet
Scissors
Overhead projector with bottom lighting
Two thin, round glass nesting bowls
White wall or sheet
Water
Water-based dye
Mineral oil
Oil-based dye
Eyedropper

STEP 1 Measure and cut the cardboard sheet so that it fits over the projector base. Cut a hole in its center that's slightly smaller than the small glass bowl.

STEP 2 Place the cardboard on the projector surface. (It will mask the bowls' edges and keep your hands from blocking the display.)

STEP 3 Position the projector so that the light shining through the cutout completely fills your target screen and the edges aren't visible.

STEP 4 Add just enough water-based dye, such as food coloring, to a glass of water to produce a light tint.

STEP 5 Place the larger glass bowl on the projector surface face up, so it's centered over the cutout. Pour the colored water in it so that its bottom is covered.

STEP 6 Combine mineral oil with oil-based dye in a separate container, and fill an eyedropper with it.

STEP 7 Drop small amounts of colored oil into the water with the eyedropper.

STEP 8 Place the small glass bowl inside the large bowl. The water and oil mixture should fill the space between the bowls.

STEP 9 To start the light show, turn on the projector (and some psychedelic tunes) and begin to move the bowls gently. Rotate them to swirl the liquids, or lift and lower them to move the image in and out of focus.

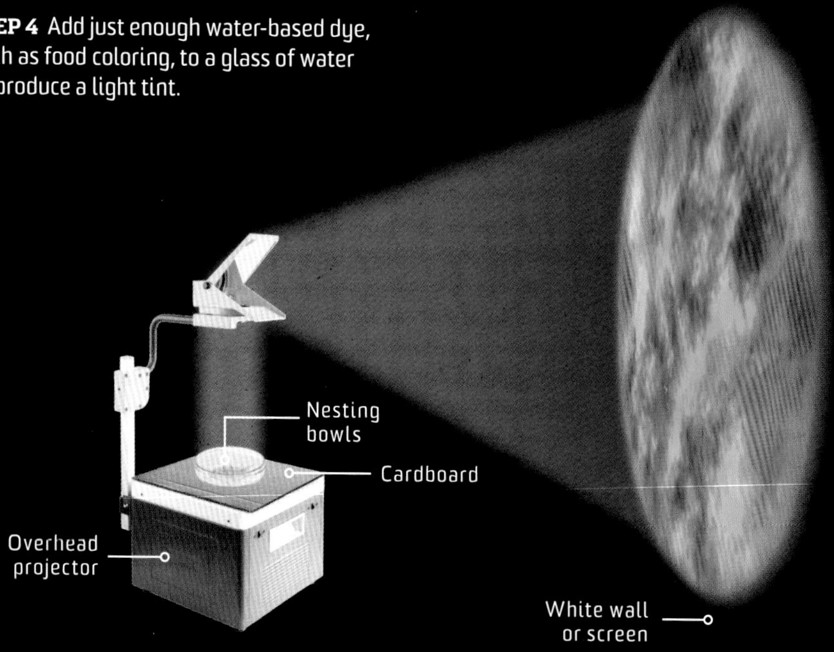

066 A HARP THAT MAKES MUSIC WITH LASERS

Ready for a sci-fi–style jam session? This floor-to-ceiling virtual instrument uses laser beams to rock for real.

Playing the harp isn't the most high-tech pastime—unless, like Stephen Hobley, you use lasers in place of the strings. Though not the first home-built laser harp, Hobley's creation is unquestionably the coolest. Played by disrupting the laser beams with the hands, it can produce just about any sound. Better yet, it's also a fully functioning controller for a version of *Guitar Hero*.

The harp consists of a box with a power supply, a 450-milliwatt green laser, a mirror, and a motherboard. After determining the beam's frequency, Hobley was able to tune a sensor so it would detect only the laser and not any ambient light. Touching a beam deflects light toward the sensor, triggering software on a PC that translates hand movements into sounds. He also wrote a script that maps notes from the harp to the keyboard controls in the videogame.

Hobley is now selling the plans for the harp on the Internet, and says he's recently had to upload a video of himself playing the game: "It was a direct response to all the comments I got to 'play *Freebird*!'"

EXTRA AMBIANCE
A fog machine placed next to the harp helps make the laser beams show up more clearly—and really adds to the mood.

067 CRAFT A BOOM BOX DUFFEL BAG

STEP 1 Create a simple image of a boom box, and draw or print it onto contact paper to create a stencil. The boom box's speakers should be more or less the same size as your speakers, which should be of the cheap desktop variety.

STEP 2 Using a craft knife, carefully cut out the stencil.

STEP 3 Lay the duffel bag flat and remove the back of the contact paper. Smooth the contact paper over the fabric.

STEP 4 Squeeze out a line of paint at the top of the stencil, and use a piece of cardboard to spread the paint over it. Repeat until the paint is well distributed. Remove the stencil.

STEP 5 After the paint has dried, cut two holes slightly smaller than your speakers out of the design. These are the holes for the speakers.

STEP 6 Remove the speakers' backs and slide the speakers into the holes from the outside. Reattach the backs with the same screws, sandwiching the fabric.

STEP 7 Plug the speakers into your portable media player, shoulder up the bag, and pump some jams.

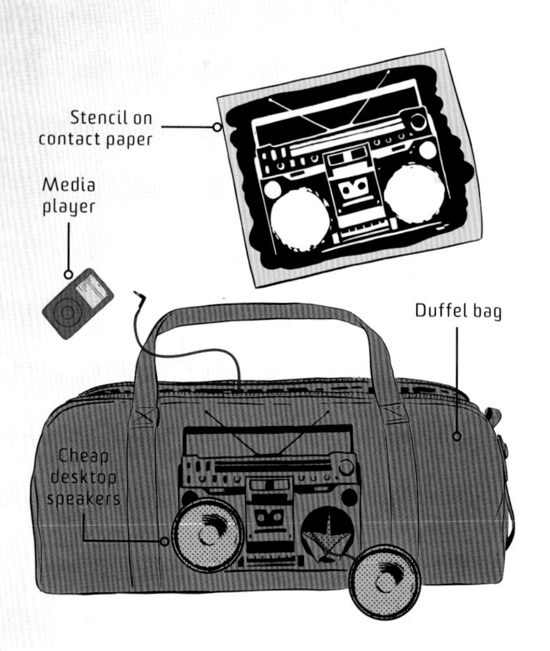

068 MAKE YOUR TIE GLOW IN THE DARK

STEP 1 Use a needle to poke a hole into the end of a tie where the EL (electroluminescent) wire will enter. It's best to use heavy-duty fabrics and to follow the seams.

STEP 2 Sew a piece of Velcro onto the tie and attach another piece to a battery pack. (We fit ours at the end of the tie.)

STEP 3 Draw a design, lay EL wire over the sketch, and tape it to the tie.

STEP 4 Measure heavy-duty fishing line that's more than twice the length of your EL wire. Thread a sturdy needle with the line and make a knot at its end.

STEP 5 Sew the wire down, securing it every 1/2 inch (1.25 cm). Remove the tape.

STEP 6 Plug the wire into the battery pack and never wear a boring tie again.

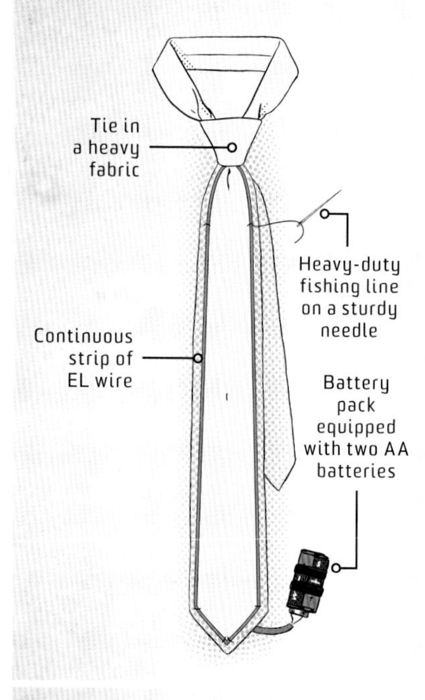

069 PUT HEADPHONES IN YOUR HOODIE

STEP 1 Use a craft knife to carefully detach the speakers from a headphone band, keeping the ear cushions and speakers (and the wire between them and the one that runs to a media player) intact.

STEP 2 Put on a lined hoodie. Safety-pin the ear cushions in place and test that they fit comfortably.

STEP 3 With a heavy-duty needle, sew four strips of Velcro onto each ear cushion's spot in the hoodie, and four more strips to the back of each ear cushion.

STEP 4 Cut a slit in the lower center of the hood where the speakers' wires will join together and enter the lining.

STEP 5 Cut a slit into the lower corner of the hoodie's front where the cord will come out.

STEP 6 Check to make sure your wire is long enough to reach the incision in the jacket's front. If it's not, desolder the cord and solder on a longer one.

STEP 7 Attach the speakers to the Velcro and feed the cord into the hood's lining and out the front.

STEP 8 Pull up your hood and get skulking. To launder, detach the ear cushions and pull out the cord.

070 USE A GLOVE ON A TOUCHSCREEN

STEP 1 Thread a sturdy needle with 1 foot (30 cm) of conductive thread.

STEP 2 On the outside of a glove's pointer finger, sew a few stitches—enough to cover an area of about 1/4 inch (6.35 mm) in diameter.

STEP 3 Turn the glove inside out, and sew three to five stitches. Allow some extra thread to dangle—this will ensure that your finger touches the conductive thread, completing a mini circuit and allowing the screen to pick up on your gestures.

STEP 4 Swipe and tap away. If you find that typing with your glove often results in hitting neighboring letters, pull out a few threads from the outside of the fingertip.

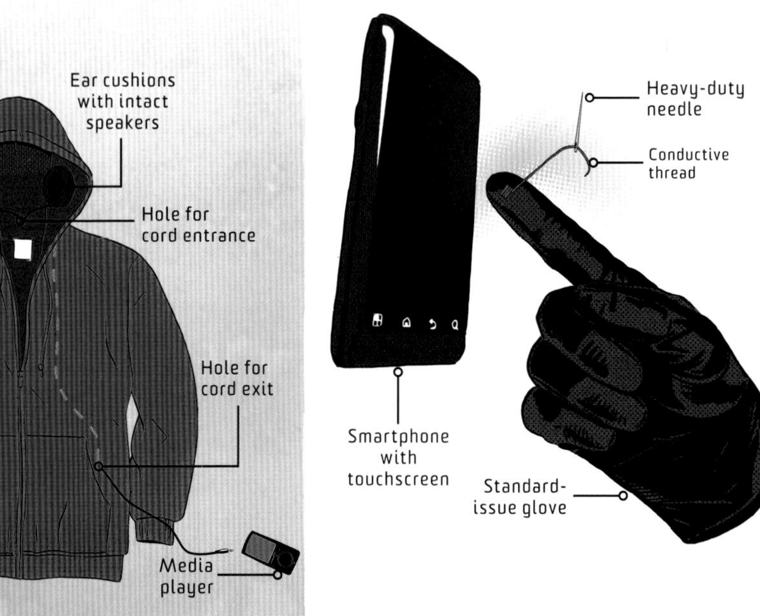

071 RIG AN ANIMATRONIC HAND

Need an extra hand? Build one—just don't blame us if it gets you in trouble.

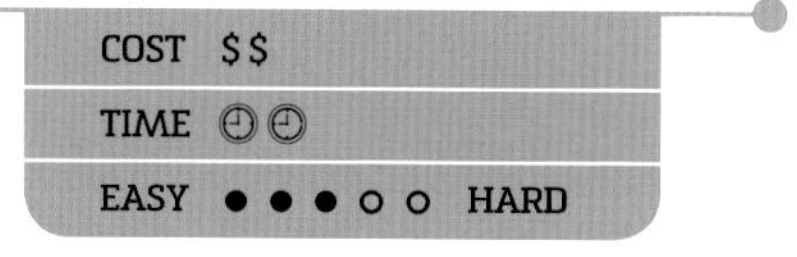

MATERIALS

Marker
Paper
Plastic tubing
Craft knife
Nylon string
Black tape
CD case
Hot-glue gun
Plastic clothes hanger
Cell foam
Glove, if desired

STEP 1 Outline your hand with a marker on a piece of paper, and mark where the joints and knuckles are.

STEP 2 Cut five "fingers" from the plastic tubing—they should be as long as the distance from your middle fingertip to your wrist.

072 MAKE A MINI WHIRLING MOTOR

Send current up over a magnetic field for some head-spinning results.

MATERIALS

6 feet (2 m) of enameled copper wire
C battery
Wire strippers
Electrical tape
Two safety pins
AA battery
Magnet

STEP 1 Wrap the wire several times around the C battery, leaving a few inches of excess at each end.

STEP 2 Slide the coil you just made off the battery. Pull one end of the excess wire through the coil, and then wrap it multiple times around the coil to hold it together. Leave about 1 inch (2.5 cm) of excess. Repeat this with the other end of the wire.

STEP 3 Strip the bits of wire extending from the coil.

STEP 4 Using electrical tape, secure safety pins to both ends of the AA battery with the hinge ends sticking up.

STEP 5 Stick the coil's stripped ends through the holes in the hinge ends so the coil is centered over the battery.

STEP 6 Place the magnet on top of the battery, then give the coil a push and watch it spin.

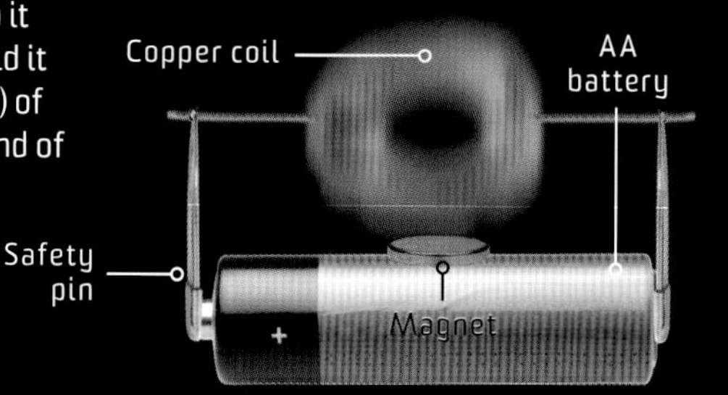

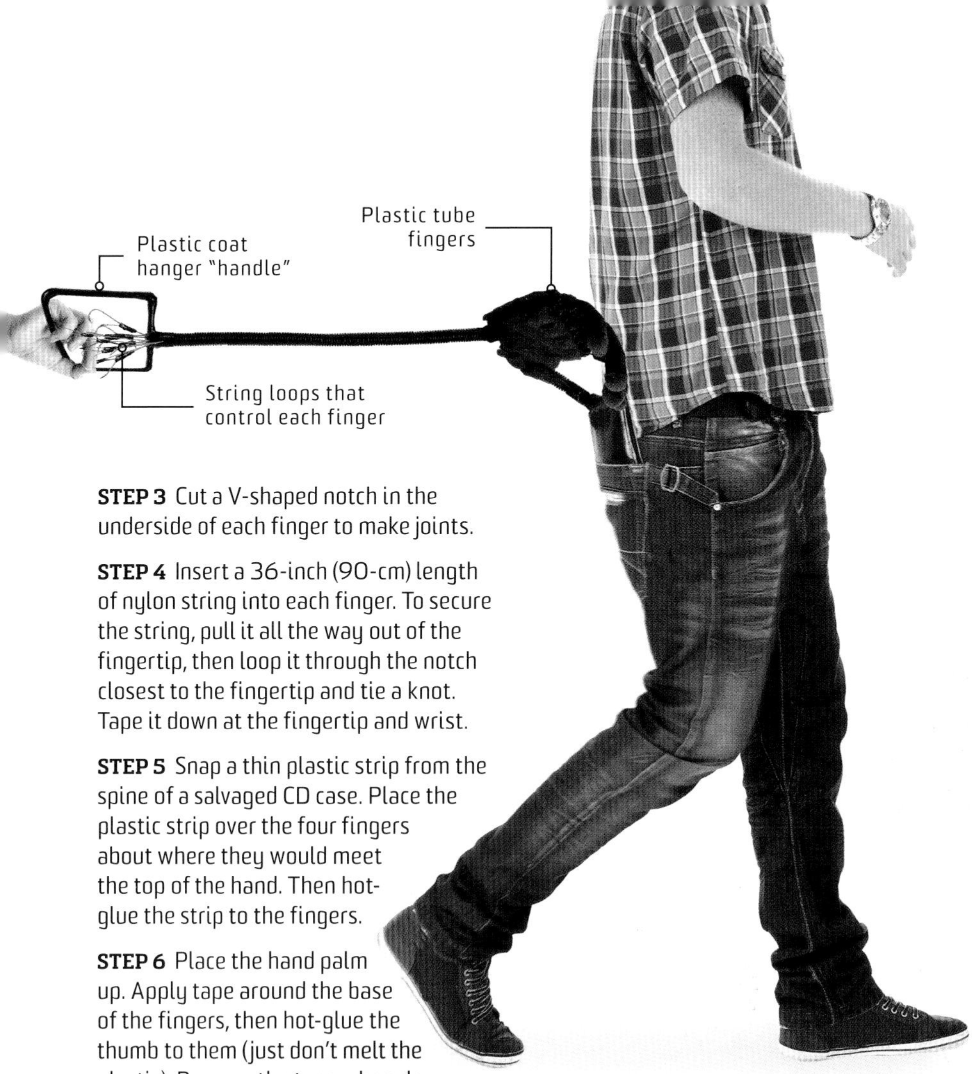

Plastic coat hanger "handle"

Plastic tube fingers

String loops that control each finger

STEP 3 Cut a V-shaped notch in the underside of each finger to make joints.

STEP 4 Insert a 36-inch (90-cm) length of nylon string into each finger. To secure the string, pull it all the way out of the fingertip, then loop it through the notch closest to the fingertip and tie a knot. Tape it down at the fingertip and wrist.

STEP 5 Snap a thin plastic strip from the spine of a salvaged CD case. Place the plastic strip over the four fingers about where they would meet the top of the hand. Then hot-glue the strip to the fingers.

STEP 6 Place the hand palm up. Apply tape around the base of the fingers, then hot-glue the thumb to them (just don't melt the plastic). Remove the tape when dry.

STEP 7 Cut another length of tubing 1 foot (30 cm) long. Thread each of the strings extending out of the wrist end of the finger tubes through the new tube.

STEP 8 Cut another piece of plastic from the CD case and glue it so it straddles the finger tubes and the arm tube, stabilizing the wrist.

STEP 9 To make a handle, cut open a plastic coat hanger and glue the pieces together into a square shape. Then glue it to the end of the arm.

STEP 10 Give your new hand the gift of human touch by applying thin pieces of foam to the fingertips and the palm of the hand. Put a glove on it if you like.

STEP 11 Make loops in the ends of all the finger strings using tape or knots. Hold one end of the square handle against the palm of your own hand, then insert your fingers into these loops and practice gently tugging on them to make the hand move creepily about.

073 MAKE A REFLECTION HOLOGRAM

Holograms aren't just for the holodeck. Make a 3D image of an object near you.

MATERIALS

Class 3A laser diode with an output of 3 to 4 mW
Tweezers
Battery pack
Wooden clothespin
Holography processing and development kit
Distilled water
A small solid object
PFG-03M 2½-by 2½-inch (6.35-by-6.35-cm) holographic plates
Cardboard
Matte black spray paint

STEP 1 Open up the laser diode and, using a pair of tweezers, remove the lens and the small tension spring. Hook the diode up to a battery pack.

STEP 2 Secure the clothespin in an upright position, and prop the laser diode between the clothespin's prongs.

STEP 3 Prepare the chemical processing solutions using distilled water and lay out and fill the trays according to the holography processing and developing kit's instructions.

STEP 4 Set up the object that you want a hologram of 15 inches (40 cm) from the laser. Glue or tape it down if you're concerned about movement.

STEP 5 Make the room fairly dark, and adjust the laser in its holder so that the beam spreads out horizontally. The object should be centered in the laser's light.

STEP 6 Place cardboard in front of the laser to block light from reaching the object.

STEP 7 In the darkest area, remove a holographic plate from its container and (after immediately closing the container) lean the holographic plate against the object so that the sticky, emulsion-coated side touches it.

STEP 8 Request that everyone in the room hold still, and lift the piece of cardboard to expose the holographic plate to the laser light for about 20 seconds. Then replace the cardboard.

STEP 9 Process the exposed holographic plate according to the holography kit's instructions, and then spray-paint the sticky side of the plate black.

STEP 10 When the plate is dry, place it in front of any incandescent (unfrosted) light source—a flashlight or the sun, for instance—to see the hologram take shape.

BUILD IT!

074 WIELD A DIY LIGHT SABER

COST	$$
TIME	◷◷
EASY	● ● ○ ○ ○ HARD

Now the force can always be with you, too.

MATERIALS

- 10-inch (25-cm) piece of 1.25 inch (3-cm) PVC pipe
- Black spray paint
- Drill
- 4 feet (1.2 m) of electrical wire
- Wire strippers
- 30 LEDs
- Pliers
- 2½-foot (75-cm) length of ¾-inch (2-cm) frosted polycarbonate tube
- Two AA batteries
- Soldering iron and solder
- On/off switch
- Duct tape

STEP 1 Spray-paint the PVC pipe black—this is your handle—then drill a hole in it for the on/off switch.

STEP 2 Cut two 2-foot (60-cm) lengths of electrical wire and strip both.

STEP 3 Attach the positive leads of the LEDs down the first wire, evenly spacing them. Then attach the LEDs' negative leads to the other stripped wire. Use pliers to crimp the LEDs' legs, securing them to the wires.

STEP 4 Drill a hole in the polycarbonate tube's top. Pull the string of LEDs into the tube through the other end, thread its wire through the hole, and tie it off.

STEP 5 Place the batteries next to each other with their polarities facing opposite directions. Attach a piece of wire across their ports with aluminum-foil duct tape.

STEP 6 Solder two wires to the on/off switch and thread it through the hole for the switch in the handle, first bringing it up through the hole in the bottom.

STEP 7 Insert the tube into the handle so that the wires from both the LED string and the switch dangle out of the handle's end. Tape the handle and tube together.

STEP 8 Solder the negative wires from the switch and LED string to the battery pack's negative port, and the positive wires from both to the positive port.

STEP 9 Duct-tape the handle's bottom to hold the battery pack inside. Turn it on and go get a Sith.

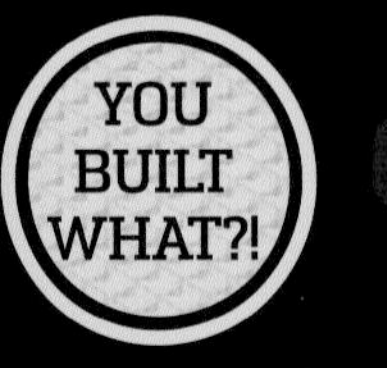

075 THE ELECTRIC GIRAFFE

It walks, it blinks, it seats six, and it blasts Kraftwerk. Meet one man's enormous pet project.

It started with a 7-inch (18-cm) walking toy giraffe and a desire to see Burning Man—the annual art-and-rave party in the desert in Black Rock, Nevada—from a higher vantage point. A year later, Lindsay Lawlor rode into the desert art festival atop Rave Raffe, a 1,700-pound (771-kg) robotic giraffe sporting 40 strobes, 400 LEDs, and bone-shaking speakers.

Lawlor wanted his Burning Man ride to be a true walking vehicle, so he copied the small toy's locomotion system on a massive scale. The front and back legs opposite each other step ahead at the same time, propelled by an electric motor. When those legs land, hydraulic brakes lock the wheeled feet, and the other two legs take a step. Canting from side to side, Raffe lumbers ahead at about 1 mile per hour (1.6 km/h). A 12-horsepower propane engine runs only to recharge the batteries, so the beast is quiet and efficient, while a pneumatic pump raises and lowers the giraffe's massive neck. When Lawlor let Raffe shuffle off alone in the desert, it walked for 8 hours.

Since the giraffe's debut, Lawlor (a part-time laser-light-show designer) has added new features, including computer-controlled flashing giraffe spots, an electroluminescent circulatory system, and a gas grill.

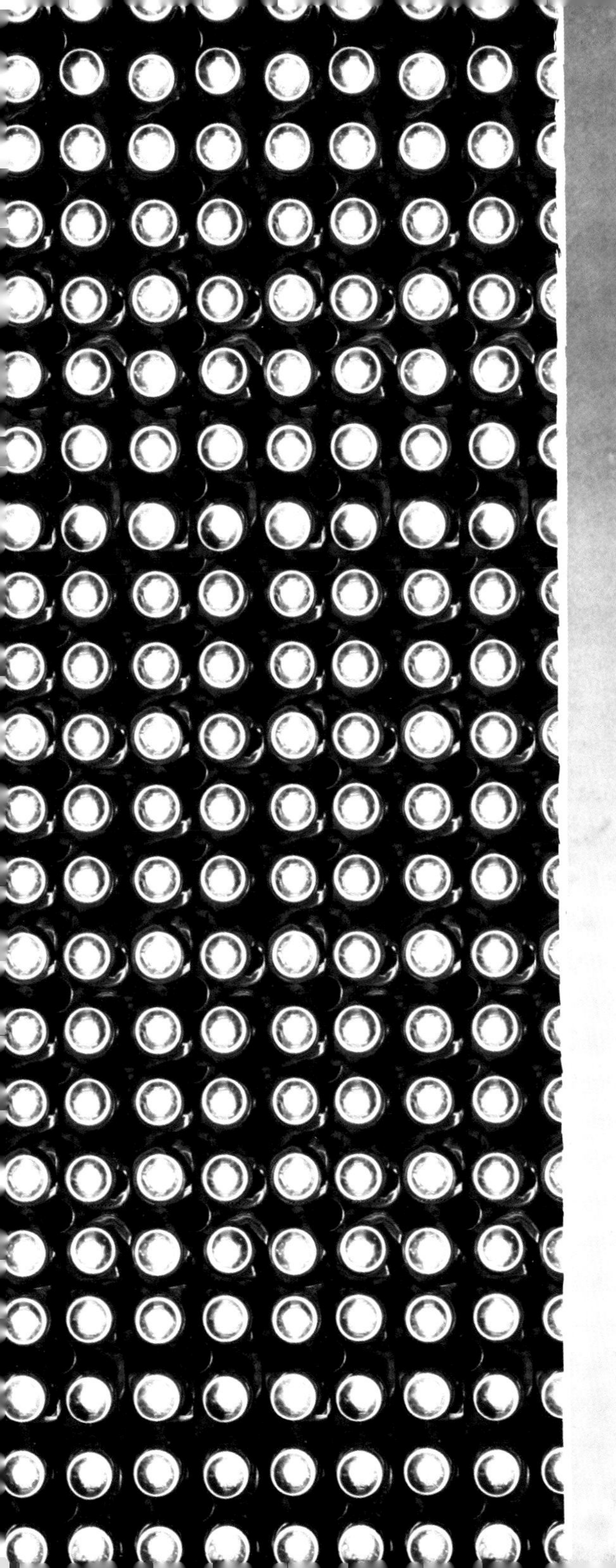

RESOURCES

GLOSSARY

ALLIGATOR CLIP Spring-loaded clip that can be used to connect a component to a wire in a temporary circuit.

ALUMINUM FLASHING Thin sheet of aluminum; often used in weatherproofing.

AMPLIFIER Component that augments the power of a signal. In circuits, an amplifier is used to increase voltage or current.

ARDUINO Common, open-source microcontroller. There are various types of Arduino microcontrollers that use the same programming language.

BLACK LIGHT Type of lamp that gives off ultraviolet light to reveal fluorescence that cannot be seen in normal lighting.

BREADBOARD Base used to set up temporary circuits before soldering them.

BREAKOUT BOARD Electrical component that allows easier access to tightly spaced pins on a microchip or bundled wires.

BUSHING Connector used to join pipes of different diameters. A bushing can also be called a reducing coupling.

CAPACITOR Electrical component that stores energy within a circuit. Unlike a battery, a capacitor does not produce energy, it simply contains or filters the energy already flowing through the circuit.

CIRCUIT Closed loop through which electrical current flows. A circuit is often used to power an electrical device.

CIRCUIT BOARD Thin, insulated board on which electrical components are mounted and connected together.

CLAMP Device used to hold an object in place. Clamps can vary widely in size and construction, and can be intended for temporary or permanent use.

COIN BATTERY Also called a button cell, a coin battery is a small, flat, disc-shaped battery often used to power portable electronic devices.

CONTACT Point where an electrical component connects to a wire or circuit board.

CONDUCTIVITY Capacity to transmit an electrical current; it can also refer to a substance's ability to transmit current.

COUPLER Short section of piping used to join two pipes.

CRAFT KNIFE Small, fixed-blade knife used to make precise cuts.

DESOLDERING Removing solder to detach components from a circuit or circuit board.

DIODE Electronic component with one terminal that has high resistance, and another terminal with low resistance. A diode is used to allow current to flow in one direction but not another.

DRILL Tool used to cut holes. A drill is usually powered by electricity, and comes with an array of interchangeable bits in different sizes.

ELECTRICAL TAPE Type of tape covered in an insulating material. Electrical tape is often used to cover and connect electrical wires.

ELECTRICAL WIRE Insulated strand of conductive material used to carry electricity.

EPOXY Adhesive made from a type of resin that becomes rigid when heated or cured.

EXHAUST FAN Fan used to ventilate a workspace; it is particularly important to use an exhaust fan when working with materials that emit toxic fumes.

FIBER OPTIC CABLE Cable made up of thin fibers that transmit light along the fiber.

FLASH DRIVE Small data-storage device that can be connected to a computer, often via a USB port.

HACKSAW Fine-toothed saw held in a frame. A hacksaw can be used to cut metal or other hard materials.

HOLE SAW Cylindrical saw blade, used with a drill to cut holes of uniform size.

HOLOGRAPHY Technique that allows for the capture of a lifelike 3D image.

INSULATION Material, such as the nonconductive coating around a wire, that prevents current or heat from flowing.

JIGSAW Tool (usually a power tool) with a long, thin saw blade. A jigsaw is useful in cutting curves and irregular shapes.

JOINT The point at which two objects are connected together.

LASER Device that emits a tightly focused beam of light. Lasers vary widely in intensity, and can be as weak as a laser pointer, or strong enough to cut through extremely hard and thick materials.

LCD MONITOR Display that uses liquid crystals and electrodes to shape light into an image.

LEAD A wire extending from an electronic component that is used to connect that component to another electronic part.

LED Diode that gives off light. They are usually much more energy-efficient than incandescent light sources, and can be much smaller.

MULTIMETER Device that measures electrical current, resistance, and voltage.

O-RING Circular seal used to join cylinders. An O-ring usually sits inside a joint to prevent leaks.

OHM Unit of measurement of electrical resistance.

PARTICLEBOARD Composite wood-based material manufactured from small chips or shavings of wood and resin.

PEG-BOARD A plank or sheet of wood pre-drilled with a grid of evenly spaced holes, Peg-Board is useful for mounting hooks and tools on a wall.

PHOTOCELL Device that produces a flow of current when exposed to light. A photocell can detect the presence (or absence) of light or other radiation.

PLEXIGLAS Hard, transparent plastic; looks like glass but is more lightweight and durable.

PORT Point of interface between one device and another. On computers, ports include Ethernet and USB ports.

POTENTIOMETER Three-terminal electrical component that acts as a variable resistor by adjusting the flow of current through a circuit.

PROGRAMMING LANGUAGE Language used to convey instructions to a computer or other machine. Many distinct programming languages are used for different purposes and types of hardware.

PROJECT BOX Box designed to contain and protect the components of a circuit.

PVC PIPE Type of durable, lightweight plastic pipe that is often used to carry liquids in plumbing.

PVC PIPE CEMENT Adhesive designed to connect pieces of PVC material together.

RESISTOR Two-terminal electrical component that resists the flow of an electric current. A resistor is used in a circuit to control the direction and strength of the current flowing through it.

ROTARY TOOL Power tool with a variety of interchangeable bits that can be used for different purposes. A rotary tool can cut, polish, carve, or grind, and is good for detail work.

SCHEMATIC Two-dimensional map of an electrical circuit. A schematic uses a set of symbols to represent the components of a circuit, and it also shows the connections that exist between these components.

SKETCH When working with Arduino microcontrollers, a sketch is a program that can be loaded into an Arduino.

SOLDERING Connecting two metal objects together by melting solder (a type of metal) with a soldering iron to create a strong joint.

SWITCH Component that can stop the flow of current in a circuit, or allow it to continue. These include push-button switches, rocker switches, toggle switches, and other devices.

TABLE SAW Machine that cuts wood or other materials with a rapidly spinning serrated metal disc. A circular saw is usually powered by electricity and mounted within a safety guard.

T-CONNECTOR Short section of pipe with three openings; used to connect lengths of pipe into a T shape.

TERMINAL End point of a conductor in a circuit; also a point at which connections can be made to a larger network.

TOUCHSCREEN An electronic display screen that users interact with by touching with their fingers; the screen detects touch within the display area.

TRANSFORMER Device that transfers current from one circuit to another. A transformer can also alter the voltage of an alternating current.

TRANSISTOR Semiconducting electrical component with at least three leads; can control or amplify the flow of electricity in a circuit.

USB Universal Serial Bus; an extremely common type of connector for computer and other electronic components.

VOLT Unit of measurement for electrical potential.

WELDING Process of joining pieces of metal together by melting them slightly and introducing a filler material at the joint.

WIRE CUTTER Pliers with sharp diagonal edges used to cut lengths of wire.

WIRE STRIPPER Device composed of a set of scissor-like blades with a central notch; used to strip the insulation from the outside of electrical wires.

INDEX

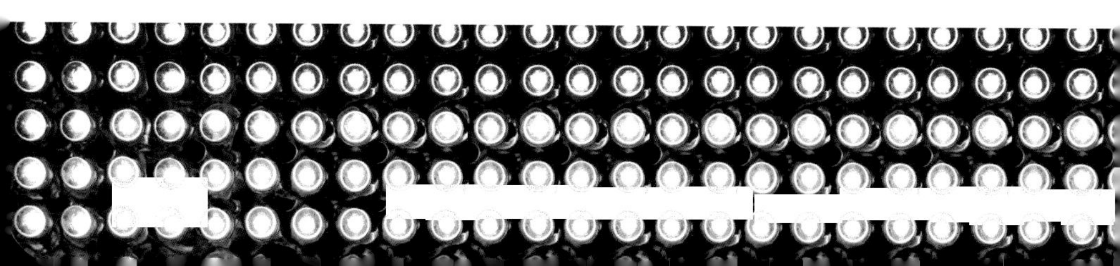

M

N

O

P

R

S

T

V

W

Z

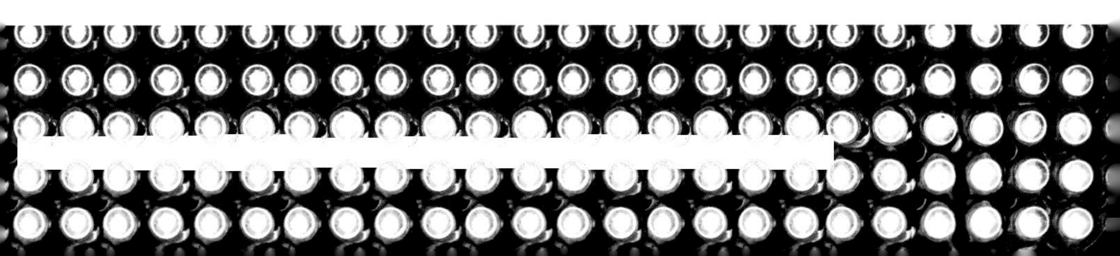

THANKS TO OUR MAKERS

Lots of inventive people contributed their ideas and how-to tutorials to the pages of this book. Look them up to find out more details about their projects, as well as any new cool stuff they're up to.

GEEK TOYS

007: John B. Carnett (carnettphoto.com) **008:** Elizabeth Hurchalla and Kent Hayward **012:** Kip Kay (kipkay.com) **014:** Jamie Price (jamiepricecreative.com) **016:** Scott McIndoe **017:** Windell H. Oskay (evilmadscientist.com) **018:** Alessandro Lambardi **019:** Kyle Pollock **020:** Vin Marshall (te-motorworks.com) **021:** Nik Vaughn **022:** Courtesy of campfiredude.com **026:** Robert Peterson (waterzooka.com) **031:** Michel Mota da Cruz **033:** Trevor Robinson **034:** Mike Andersen, Grant Elliot, Schyler Senft-Grup, and Scott Torborg (scotttorborg.com) **035:** Michael Greensmith (steampunkwayoflife.blogspot.com.au) **037:** Maayan Migdal (created at the Bezalel Academy of Art and Design, under professors Ytai Ben-Tsvi, Shachar Geiger, and Itay Galim) **039:** Aram Bartholl (deaddrops.com) **041:** Dave Prochnow **043:** Inspired by Kang Chang, Kyle Milns, and Mike Fleming; further developed by Ian Cannon **044:** Courtesy of illphabetik.com **045:** Inspired by a tutorial by Instructables username hunrichs; furthered by Emelie Griffin **046:** Kimanh le Roux (scissorspaperwok.com) **047:** Emelie Griffin **048:** Spudtech (spudtech.com) **049:** Anthony Le (masterle247.wix.com) **055:** Eddie Zarick **057:** Joshua Zimmerman (browndoggadgets.com) **059:** Harout Markarian **060:** Daniel Wolf (cookrookery.com) **061:** Jason Wilson **062:** Robert Waters **064:** Andrew Lim (cofounder of Recombu.com) **065:** R. Lee Kennedy, Associate Professor, Department of Drama, University of Virginia **066:** Stephen Hobley (stephenhobley.com) **070:** Tim Lillis **071:** Bard Lund Johansen **073:** A.C. Jeong **074:** Michael Nagle **075:** Lindsay Lawlor (electricgiraffe.com)

IMAGE CREDITS

All images courtesy of Shutterstock Images unless otherwise noted.

Courtesy of the *Popular Science* Archives: 028, 054, 058 **Conor Buckley:** 008, 015, 017–019, 021 (Rubens' tube illustration), 026–027, 029, 030–33 (lighter illustration), 036–039, 040, 042–048, 050, 052–053, 056–057, 063–065, 071–074 **John B. Carnett:** 007, 014, 020–021, 034 **Getty Images:** 041 (Léon Theremin photograph) **Michael Greensmith:** 035 **JP Greenwood:** 075 **Stephen Hobley:** 066 **Scott McIndoe:** 016 **Tyler Stableford:** 049 **Nik Vaughn:** 021 (Rubens' tube photograph) **Carl Wiens:** 009–012, 022–025, 059–062, 067–070 **Eddie Zarick:** 055

DISCLAIMER

The information in this book is presented for an adult audience and for entertainment value only. While every piece of advice in this book has been fact-checked and where possible, field-tested, much of this information is speculative and situation-dependent. The publisher assumes no responsibility for any errors or omissions and makes no warranty, express or implied, that the information included in this book is appropriate for every individual, situation, or purpose. Before attempting any activity outlined in these pages, make sure you are aware of your own limitations and have adequately researched all applicable risks. This book is not intended to replace professional advice from experts in electronics, woodworking, metalworking, or any other field. Always follow all manufacturers' instructions when using the equipment featured in this book. If the manufacturer of your equipment does not recommend use of the equipment in the fashion depicted in these pages, you should comply with the manufacturer's recommendations. You assume the risk and full responsibility for all of your actions, and the publishers will not be held responsible for any loss or damage of any sort—whether consequential, incidental, special, or otherwise—that may result from the information presented here. Otherwise, have fun.

weldon**owen**

President, CEO Terry Newell
VP, Sales Amy Kaneko
VP, Publisher Roger Shaw
Senior Editor Lucie Parker
Project Editors Emelie Griffin, Jess Hemerly
Creative Director Kelly Booth
Designer Michel Gadwa
Image Coordinator Conor Buckley
Production Director Chris Hemesath
Production Manager Michelle Duggan

415 Jackson Street, Suite 200
San Francisco, CA 94111
Telephone: 415 291 0100
Fax: 415 291 8841
www.weldonowen.com

Popular Science and Weldon Owen are divisions of **BONNIER**

Library of Congress Control Number is on file with the publisher.

ISBN 13: 978-1-61628-490-9
ISBN 10: 1-61628-490-0

10 9 8 7 6 5 4 3 2 1
2013 2014 2015

Printed in China by 1010 Printing.

POPULAR SCIENCE THE FUTURE NOW

ACKNOWLEDGMENTS

Weldon Owen would like to thank Jacqueline Aaron, Katie Cagenee, Andrew Jordon, Katharine Moore, Gail Nelson-Bonebrake, Jenna Rosenthal, Katie Schlossberg, and Marisa Solís for their editorial expertise and design assistance.

We'd also like to thank our technical editors, Michael Rigsby and Tim Lillis, and our in-house builder and circuity diagram consultant, Ian Cannon.

Popular Science would like to thank Matt Cokeley, Todd Detwiler, Kristine LaManna, Stephanie O'Hara, Thom Payne, and Katie Peek for their support over the years.

We would also like to thank Gregory Mone for penning the You Built What?! entries included in this book.

And a big thanks to Mark Jannot and Mike Haney—the How 2.0 column's first editor—for getting it all started.